THINKING STRATEGIES
FOR
SOLVING PROBLEMS

THINKING STRATEGIES
FOR
SOLVING PROBLEMS

MAILOO SELVARATNAM
PhD (Lond.); D.I.C; F.R.S.C; C.CHEM.

Formerly Professor of Chemistry at North-West University, Mafikeng Campus,
South Africa (1985-2013) and University of Peradeniya, Sri Lanka (1960-1984)
Email: mailooselva@gmail.com

PARTRIDGE

Copyright © 2016 by Mailoo Selvaratnam.

ISBN:	Softcover	978-1-4828-6276-8
	eBook	978-1-4828-6277-5

All rights reserved. No part of this book may be used or reproduced by any means, graphic, electronic, or mechanical, including photocopying, recording, taping or by any information storage retrieval system without the written permission of the author except in the case of brief quotations embodied in critical articles and reviews.

Because of the dynamic nature of the Internet, any web addresses or links contained in this book may have changed since publication and may no longer be valid. The views expressed in this work are solely those of the author and do not necessarily reflect the views of the publisher, and the publisher hereby disclaims any responsibility for them.

Print information available on the last page.

To order additional copies of this book, contact
Toll Free 0800 990 914 (South Africa)
+44 20 3014 3997 (outside South Africa)
orders.africa@partridgepublishing.com

www.partridgepublishing.com/africa

Contents

PREFACE 1

CHAPTER 1
INTRODUCTION 3

CHAPTER 2
CLARIFICATION AND CLEAR
PRESENTATION OF PROBLEMS 7

CHAPTER 3
IDENTIFICATION AND FOCUSING ON THE GOAL 34

CHAPTER 4
IDENTIFICATION OF PRINCIPLES
NEEDED FOR SOLUTIONS 49

CHAPTER 5
USE OF EQUATIONS FOR
CALCULATIONS AND DEDUCTIONS 76

CHAPTER 6
PROCEEDING STEP - BY – STEP WITH
THE SOLUTION 103

PREFACE

This monograph is intended primarily for lecturers/instructors of high school and first year university science courses, and for motivated science students at universities, tertiary institutions and high schools. Parts of the book may also be used by persons without a science background because many problems in the monograph do not need science concepts for their solutions. The non-science problems (examples and exercises) are given first in each chapter, before the science problems, to assist readers without a science background.

The monograph discusses cognitive strategies (thinking strategies) that are fundamentally important for solving problems encountered in physical science courses. Many of these strategies are also important for solving problems encountered in our daily lives. The emphasis throughout the monograph is on the strategies one should use for solving problems efficiently, on how one should set-about solving a problem, and not just on obtaining the solution.

Cognitive strategies and skills are important because they are the "tools" for all types of learning and problem solving. They are also an important factor that affects our intelligence (Intelligence Quotient, IQ) and self-confidence.

An important objective of all academic courses should therefore be the training of students in cognitive skills and strategies

A large amount of research evidence suggests that the learning difficulties of many students are due to their not being sufficiently competent in cognitive strategies and skills. This may be one of the important reasons for "rote-learning" by many students.

The Examples and Exercises in this monograph are intended to be used by teachers to train students. Training in cognitive strategies and skills is best done by *integrating* them, at the appropriate places, with the teaching of subject-content knowledge. Repeated training throughout a course will be needed to ensure that students become competent in them. The use of these strategies should become a habit of the mind.

I am particularly grateful to the following who contributed significantly to the improvement of the monograph: Mr. Placid Fernandez, St. Michaels High School, Mafikeng; Ms. Millicent Mongale, North-West University, Mafikeng Campus; Prof. Sebastian Canagaratna, Ohio Northern University, USA. Others whom I should thank for helping me in the preparation of the monograph are: Prof. Helen Drummond and Ms. Mathola Kefilwe from North-West University, my wife Hema, my daughter in law Wathsala Dissanayake and Mr. Thayendran Naidoo.

CHAPTER 1
INTRODUCTION

An important objective of academic courses should be the improvement of the cognitive abilities (thinking abilities) of students because they are the "tools" for all types of learning, decision-making and problem-solving. Competence in cognitive abilities is hence essential if students are to learn efficiently and also solve problems they encounter, not only in their academic courses but also throughout their lives.

Increased competence in cognitive abilities can also be expected to increase students' self-confidence and develop their intelligence. Research suggests that intelligence (Intelligence Quotient, IQ) depends not only on genetic inheritance but also on ones level of competence in cognitive abilities (particularly cognitive strategies).

Intelligence can be improved by providing explicit training in cognitive strategies and skills.

There are many types of cognitive abilities and they have been classified in different ways and at different depths of detail. A broad classification is as cognitive skills (thinking skills) and cognitive strategies (thinking strategies).

Cognitive skills are the basic building-blocks of all mental activities. Some of them are of general applicability (e.g. skills associated with reading, writing and dealing with numbers) while others are specialized. Specialized skills are often needed for the effective learning of science and this includes mathematical skills, organization skills, information-processing skills, three-dimensional visualization skills and various types of reasoning skills (e.g. deductive reasoning, inductive reasoning, and analogical reasoning).

Cognitive strategies are plans of action for guiding, controlling and executing mental tasks associated with learning and use of knowledge. They determine how we set-about learning a topic or approach the solution of a problem. Cognitive strategies are generally linked with cognitive skills and the execution of a strategy often needs competence in some cognitive skills. Unlike a cognitive skill, which is often narrow and specific to subject-content, a cognitive strategy is broad and is generally applicable to a variety of situations and contexts. Just a few cognitive strategies are needed for effective learning and problem solving, in contrast to cognitive skills where many are often needed.

Extensive research has shown that many students do not have sufficient competence in the cognitive strategies and skills needed for effective learning and problem solving. This lack of competence will seriously handicap effective learning and problem solving, and may be an important reason for "rote-learning" (memorization without much understanding) by many students.

Thinking Strategies for Solving Problems

The main purpose of this monograph is to consider and discuss some cognitive strategies that are important for effective problem solving in science courses. Five important cognitive strategies are:

- Clarification and clear representation of problems;
- Identification of the goal and formulating an approach for reaching the goal;
- Identification and use of the principles needed for the solutions;
- Use of equations for calculations and deductions;
- Proceeding step-by-step with the solutions.

These strategies are often not independent and they are related to one another. Some of the examples given under a particular strategy may equally well have been given under another strategy. For example, some of the examples given under the strategy "use of equations" could also have been given under the strategy "use of principles". This is because equations summarize principles.

These strategies are not difficult to understand and use. Many students however do not use them. This is mainly because sufficient emphasis is not placed on them in their courses. Repeated training in them is needed to ensure that the use of these strategies becomes an automatic habit of the mind.

Testing ones competence in a thinking strategy using questions (in a written test) is often difficult. This is because the difficulty in answering a question must be due only to incompetence in the use of a strategy and not to any other reason such as difficulty with subject-content knowledge or language. To ensure that lack of subject-content knowledge does not cause much difficulty, simple questions have therefore to be designed: questions that do not need difficult science concepts and principles for their solutions.

The concepts/ principles required for the solutions of the examples/ exercises given in this monograph are simple and are:

Speed, density, concentration, ideal gas equation, law of conservation of mass, balanced equations.

To obtain the greatest benefit from the monograph you should try to solve the examples on your own, before reading through the solutions. This will help you to obtain a greater appreciation of the importance and usefulness of the strategies for successful problem solving.

CHAPTER 2
CLARIFICATION AND CLEAR PRESENTATION OF PROBLEMS

VERBAL INFORMATION	VISUAL INFORMATION
On a particular day, when it is 12 h in Johannesburg (JHB) it is 6 h in New York (NY), and when it is 16 hours in Sydney (SYD) it is 8 h in JHB.	6h 8h NY JHB SYD

Which type of information (verbal or visual) is easier to use for solving the problem:

What will be the time in NY when it is 18 h in SYD?

A clear presentation of a problem will often simplify problem solving and help avoid errors. Many errors and difficulties of students during problem solving arise because they rush-in into the solution without placing sufficient emphasis on this step. This is a broad strategy that involves many aspects and it includes the following:

- Identifying all the quantities (data, goal) involved in the problem and distinguishing them by giving different symbols.

- Arranging the data, the goal and any processes involved in one place, in a clear, concise, coherent and coordinated manner, for example as

    - diagrams
    - graphs
    - tables
    - equations

Clarification of a problem is not just an initial step during problem solving. It has often to be done continuously throughout the various steps in the problem-solving process.

The importance of using these strategies will now be illustrated with examples, some of which have been tested with many groups of first year university students. You should try to solve the examples on your own before reading through the given solutions. This will help you to obtain a greater appreciation of the importance of the strategy. To

Thinking Strategies for Solving Problems

encourage you to do this, the questions in the examples are given below. The solutions to them are given later.

2.1. Consider three objects A, B and C. The mass of A (symbol m_A) is larger than the mass of B (symbol m_B) by 6 grams but is smaller than the mass of C (symbol m_C) by 8 grams. If the mass of C is 20 grams, calculate the mass of B.

2.2. This question concerns corresponding times, on a particular day, in Johannesburg (JHB), Sydney (SYD) and New York (NY). When it is 12 h in JHB, it is 6 h in NY and when it is 16 h in SYD it is 8 h in JHB. What will be the time in NY when it is 18 h in SYD?

2.3. Consider the two statements: All mammals are warm-blooded animals; Animal A is warm-blooded. Can we conclude from these two statements that A is a mammal? Indicate your reasoning.

2.4. A solid substance melts at $-40°$ C to form a liquid which boils at $80°$ C to form a gas. State whether this substance will be a solid, liquid or gas when its temperature is:
(a) 50°C,   (b) 90°C,   (c) - 50°C

2.5. Two scales for measuring temperature are the Celsius scale ($°C$) and the Fahrenheit scale ($°F$). Given that $0°$ C = $32°$ F and $100°$ C = 212 $°F$, calculate the temperature in $°C$ when it is $122°$ F.

2.6. A closed vessel at 25° C contains 0.10 mole of nitrogen, 0.15 mole of hydrogen and 0.50 mole of ammonia. When this gaseous mixture is heated to 300° C, 0.20 mole of ammonia dissociates according to the equation $2NH_3$ (g) $\rightarrow$ N_2 (g) $+3H_2$ (g). Nine types of amounts (n) are associated with this problem. Identify them and give different symbols to distinguish them.

2.7. EXAMPLE 2.1

Consider three objects A, B and C. The mass of A (symbol, m_A) is larger than the mass of B (symbol, m_B) by 6 grams but is smaller than the mass of C (symbol, m_C) by 8 grams. If the mass of C is 20 grams, calculate the mass of B.

Solution

The solution is easy if the data are first represented *visually* on a line diagram, as shown below. The visual representation of a problem always simplifies its solution.

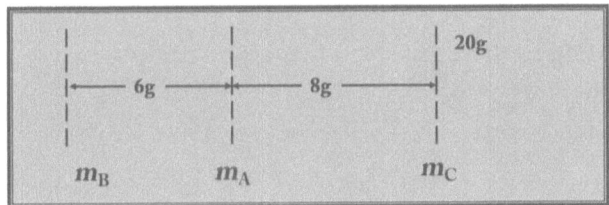

This diagram shows that m_B is less than m_C by 14 g (8 g + 6 g) and hence

$$m_B = 20 \text{ g} - 14 \text{ g} = 6 \text{ g}$$

Another method of correlating the data, since the data given are quantitative, is to represent them as equations

$$m_A = m_B + 6 \text{ g}$$
$$m_A = m_C - 8 \text{ g}$$
$$m_C = 20 \text{ g}$$

From these equations the mass of B can easily be calculated.

Tests with first year university students showed that about a third of them had difficulty with this problem. Their difficulties were mainly associated with their not initially correlating the given data in a line diagram or as equations. Tests with other questions also showed that students' difficulties were of two types:

- Insufficient awareness of the importance of this strategy;
- Inability to represent the information given in this problem in a line or as equations.

Training in these aspects is hence needed.

EXAMPLE 2.2

This question concerns corresponding times, on a particular day, in Johannesburg (JHB), Sydney (SYD) and New York (NY). When it is 12 h in JHB, it is 6 h in NY and when it is 16 h in SYD it is 8 h in JHB. What will be the time in NY when it is 18 h in SYD?

Solution

The solution will be simplified if the given data and the goal are first presented together in a Table (see rows 1, 2 and 3 in the table below).

Collection of all the relevant information in one place will help us to focus sharply on the solution, without being distracted by another step (the simultaneous search for the needed information).

The steps in the solution are also shown in the Table in rows 4, 5 and 6 which shows that 18 h in SYD = 4 h in NY. Tabulation of the steps will guide us with the solution.

	SYD	JHB	NY	
		12 h	6 h	
Data	16 h	8 h		
Goal	18 h		? h	
Solution	18 h	10 h		Add 2 h to data in row 2
		10 h	4 h	Subtract 2 h from data in row 1
	18 h	10 h	4 h	Combine information in last 2 rows

The solution will be simplified even further if the *difference* between the times in NY, JHB and SYD is shown visually in a line diagram, which shows clearly that when it is 18 h in SYD it will be (18 - 8 - 6) h = 4 h in NY.

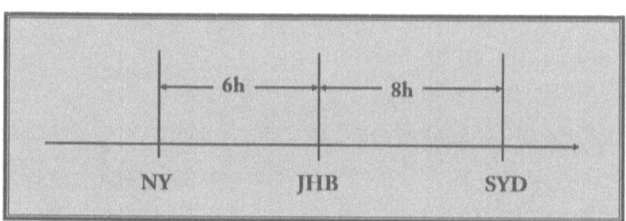

Tests with first year university students showed that about a third of them could not solve this problem. Their difficulty was mainly due to their not recognizing the importance of initially correlating the given data in a Table or in a line diagram.

EXAMPLE 2.3

Consider the two statements: All mammals are warm-blooded animals; Animal A is warm-blooded.

Would it be correct to conclude from these two statements that A is a mammal? Indicate your reasoning.

Solution

The solution will be simplified if all the given information is first presented *visually* in a diagram (called "Venn" diagram). As a general strategy, the information given in a problem should be represented, whenever possible, as a diagram. This will aid solution of the problem.

The information in the first statement which involves two concepts (mammal, warm-blooded) is shown in fig.(a) where the larger rectangle represents warm-blooded animals and the smaller rectangle the information that " all mammals are warm-blooded animals". Figure (b) shows the diagram obtained when the second statement "Animal A is warm-blooded" is included in fig (a). It shows that A can be both inside and outside the smaller rectangle. This figure shows that A could be a mammal but need not be a mammal (since A could also be outside the smaller rectangle that represents mammals).

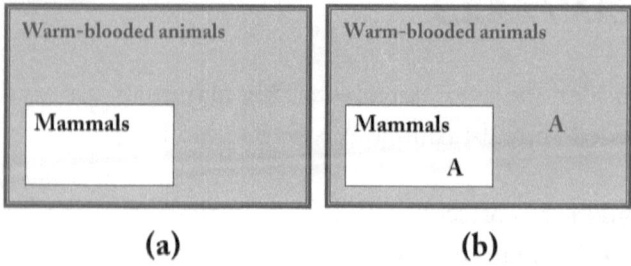

(a)   **(b)**

It would therefore be incorrect to conclude that A is a mammal.

Tests with first year university students showed that about a half of them had difficulty. Their difficulty was mainly due to their trying to use *verbal reasoning* to solve the problem.

Thinking Strategies for Solving Problems

EXAMPLE 2.4

A solid substance melts at - 40°C to form a liquid which boils at 80°C to form a gas. State whether this substance will be a solid, liquid or gas when its temperature is

(a) 50°C   (b) 90°C   (c) -50°C

Solution

The solution is easy if all the data given are first correlated together in one place, for example in a *line diagram* as shown below.

The solution then merely needs ability to read the information given in this diagram, which shows that the substance will be a liquid at 50°C, a gas at 90°C and a solid at -50°C. *No reasoning is needed to answer the question.*

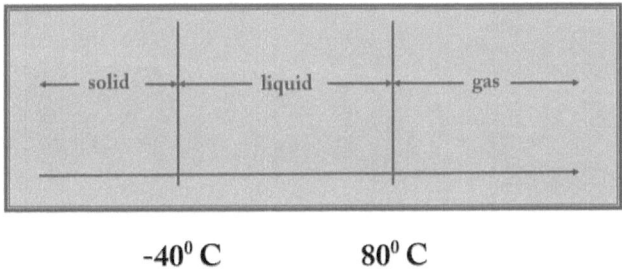

This question has been tested with many groups of first year university students and it was found that about a third of them had difficulty. The main reason for the difficulty was associated with their not correlating all the data in one place. Some students had difficulty with one or two parts of the question which suggests that they solved each part separately. This would need more time and is undesirable. It will be much better to show all the data as a *line diagram* and use this diagram to answer all parts of the question.

EXAMPLE 2.5

Two scales for measuring temperature are the Celsius scale (°C) and the Fahrenheit scale (°F). Given that 0°C = 32°F and 100°C = 212°F, calculate the temperature in °C when it is 122°F.

Solution

Correlating the data in a diagram will assist solution. The following diagram shows the correspondence between the given data (0°C = 32°F, 100°C = 212°F) and also indicates the centigrade temperature corresponding to 122°F as x

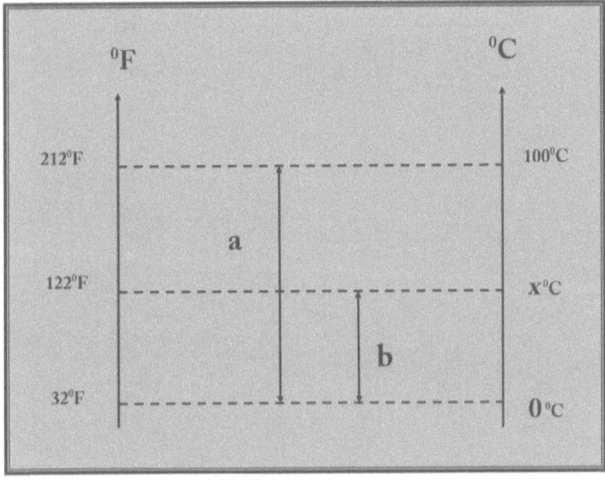

From the diagram it can be seen that the ratio a/b is equal to (212 -32)°F/ (122-32)°F on the Fahrenheit scale and to (100 - 0)°C / (x- 0)° C on the Celsius scale. Hence we can write

$$\frac{(212 - 32)\,°F}{(122 - 32)\,°F} = \frac{(100 - 0)\,°C}{(x - 0)\,°C}$$

Simplification of this equation will show that $x = 50$. The temperature corresponding to $122°F$ is therefore $50°C$.

Note that it is also possible to deduce the answer from the diagram by direct- proportion reasoning.

The diagram shows that $(212 - 32)°\,F = (100 - 0)°\,C$. Therefore $(122 - 32)°\,F$, *by direct proportion reasoning*, will be equal to

$$\frac{(100-0)\,°C}{(212-32)\,°F} \times (122-32)\,°F = 50°C$$

EXAMPLE 2.6

A closed vessel at $25°C$ contains 0.10 mole of nitrogen, 0.15 mole of hydrogen and 0.50 mole of ammonia. When this gas mixture is heated to $300°C$, 0.20 mole of ammonia dissociates according to the equation $2NH_3(g) \rightarrow N_2(g) + 3H_2(g)$. Nine types of amounts (n) are associated with this problem. Identify them and give different symbols to distinguish them.

Solution

Three substances (N_2, H_2 and NH_3) are present in the vessel. Amounts (moles) of them can be distinguished by giving the symbols n_{N2}, n_{H2} and n_{NH3}. Since a chemical reaction takes place, it is also necessary to distinguish between initial amounts, amounts reacted and amounts produced, and amounts finally present in the vessel. Nine types of amounts are associated with this problem (see left column of table) and explicit symbols for them are given in the right column.

AMOUNT	SYMBOL
initial amount of N_2	n_{N2} (init.)
initial amount of H_2	n_{H2} (init.)
initial amount of NH_3	n_{NH3} (init.)
amount of NH_3 dissociated	n_{NH3} (diss.)
amount of H_2 produced by dissociation	n_{H2} (prod.)
amount of N_2 produced by dissociation	n_{N2} (prod.)
amount of N_2 finally present in vessel	n_{N2} (final)
amount of H_2 finally present in vessel	n_{H2} (final)
amount of NH_3 finally present in vessel	n_{NH3} (final)

Mailoo Selvaratnam

Many difficulties and errors of students are due to their not distinguishing the different quantities involved in a problem, and then not giving different symbols to distinguish the different quantities.

Thinking Strategies for Solving Problems

EXERCISES

Exercise 2.1
This problem concerns the masses of five objects A, B, C, D and E. Object A is heavier than B but lighter than C. Object D is lighter than A but heavier than B. Object E is lighter than B. Arrange the objects in the order of increasing mass.

Exercise 2.2
From a town A, a car travels 5 km east (E), then 3 km north (N), then 10 km west (W), then 6 km south (S) and finally 5 km east (E) to reach town B. What is the distance between towns A and B?

Exercise 2.3
A boat rows north (N) at a speed of 4 kilometres per hour across a river which flows east (E) at a speed of 2 kilometres per hour. Represent these speeds, roughly to scale, as a diagram. Then draw a line in this diagram to indicate the direction in which the boat will move.

Exercise 2.4
Three stationary trains A, B and C are parked on different railway lines that are parallel to one another. The front end of train A is 30 metres (m) in front of the front end of train B and the rear end of train B is 10 m behind the rear end

of train A. The front end of train C is 10 m in front of the front end of train B and the rear end of train C is in the same vertical line as the rear end of train A. If the length of train C is 50 m, what are the lengths of trains A and B?

Exercise 2.5
Tanyia had a chocolate. On the first day she ate one quarter of it and on the second day Arjuna ate a third of the remaining chocolate. On the third day Onali ate a half of the chocolate that was left. What fraction of the chocolate would then remain uneaten, for Nushara to eat.

Exercise 2.6
Consider four objects A, B, C and D. The mass of B is larger than the mass of C by 4 grams but smaller than the mass of A by 8 grams. The mass of D is larger than the mass of C by 2 grams. If the mass of A is 20 grams, what will be masses of B, C and D?

Exercise 2.7
Consider the two statements: All metallic substances are electrical conductors; Substance A is an electrical conductor. Would it be logical to conclude from these two statements that A is a metal? Explain.

Exercise 2.8
Consider the two statements: Some animals are warm-blooded. A is not warm-blooded. Can we logically conclude from these two statements that A is not an animal? (Hint: draw a "Venn diagram").

Thinking Strategies for Solving Problems

Exercise 2.9
Every hour (e.g. at 4 am, 5 am, 6 am ...) a train leaves railway station A to travel to station B and a train leaves B to travel to A. Both trains travel at a constant speed of 50 km per hour and the time taken for travel from A to B (and from B to A) is 3 hours. Consider the train that leaves A at 8 am. How many trains, travelling from B to A, will this train meet when it travels towards B? Find also the times of the meetings. (Assume that the speeds of the trains are 50 km per hour even when they start/ stop their journey).

Exercise 2.10
1.0 mole of alcohol (a liquid) is dissolved in 2.0 litres of water. The concentration (c) of alcohol in the solution then obtained will be (select the correct answer; $c = n/V$ where n = number of moles, V = volume)

(a) 0.50 mol l^{-1} (b) less than 0.50 mol l^{-1} (c) greater than 0.50 mol l^{-1} (d) 1.0 mol l^{-1}

Exercise 2.11
5 grams of a substance A are present in a closed 2 litre vessel at $20°C$. When it is heated to $300°C$, it partly breaks down to give 1 gram of B and 0.8 gram of C. Calculate the mass of A then present in the vessel by applying the law of conservation of mass which states that the total mass does not change during any reaction.

Exercise 2.12
Consider an *aqueous* solution of hydrochloric acid (HCl), at $25°C$, whose concentration (c) is 1×10^{-5} M. Since HCl

is fully dissociated, c_{H^+} will be equal to 1×10^{-5} M. If this solution is diluted by a factor of 10 (e.g. by diluting 1 cm³ of this HCl solution to 10 cm³ by adding water), c_H^+ will decrease by a factor of 10 and hence will be 1×10^{-6} M. What will be c_H^+ if the 1×10^{-5} M HCl solution is diluted by a factor of 1000?

Did you get 1×10^{-8} M as the answer? This cannot be correct because an acidic solution ($c_H^+ > 10^{-7}$ M is acidic) cannot become a basic solution ($c_H^+ < 10^{-7}$ M is basic) by diluting it with water. Identify the source of error in your calculation.

Thinking Strategies for Solving Problems

SOLUTIONS TO EXERCISES

Exercise 2.1
Correlate all the information given in one place (e.g. in a line diagram). The order will then be seen to be: E B D A C.

Exercise 2.2
A diagram showing the steps in the travel will simplify the solution. The diagram below shows that the distance between towns A and B is 6 km - 3 km = 3 km.

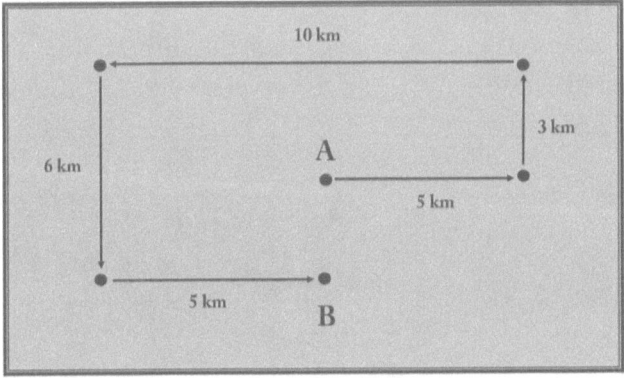

Exercise 2.3

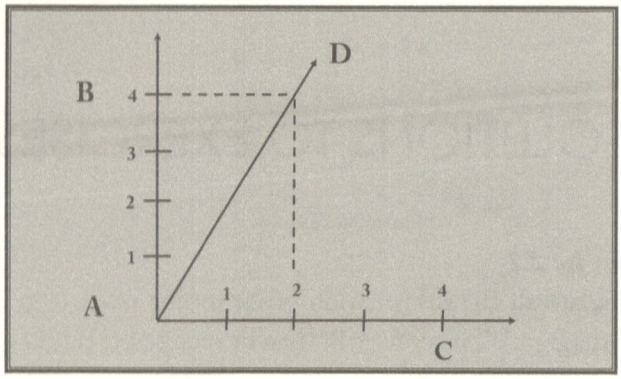

In the diagram above, 4 and 2 represent respectively the speeds in the North and East directions and the line AD shows the direction in which the boat will move.

Exercise 2.4
The solution will be simplified if the information given about the trains A, B and C are presented together in a diagram.

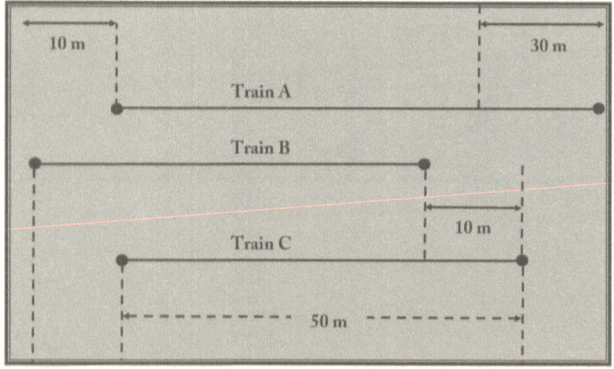

Thinking Strategies for Solving Problems

From the diagram it can be seen that
Length of train B = length of train C - 10m + 10m = 50 m
Length of train A = length of train B - 10 m + 30 m = 70 m

Exercise 2.5

Eaten 1st Day	Eaten 2nd Day	Eaten 3rd Day	Uneaten

The diagram above simplifies the solution. It divides the chocolate into four equal parts and represents all the information given in the problem. It shows that the fraction of the chocolate remaining uneaten is 1/4.

Exercise 2.6

The solution will be simplified by correlating all the information given, for example in a line diagram as shown below.

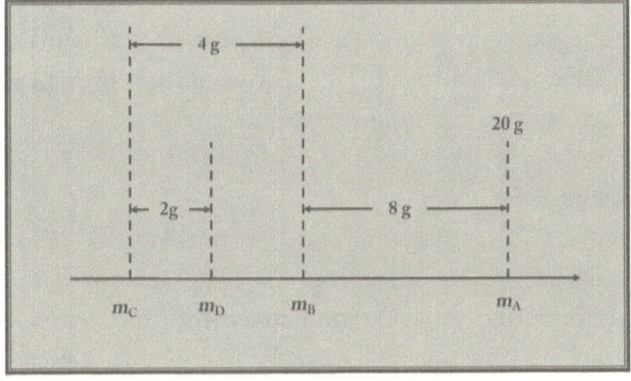

From the diagram it is easy to see that
$$m_B = 20 \text{ g} - 8 \text{ g} = 12 \text{ g}$$
$$m_C = 20 \text{ g} - 8 \text{ g} - 4 \text{ g} = 8 \text{ g}$$
$$m_D = m_C + 2 \text{ g} = 8 \text{ g} + 2 \text{ g} = 10 \text{ g}$$

Exercise 2.7

Reasoning to obtain solution will be made easier by presenting the information given visually in a diagram (see example 2.3)

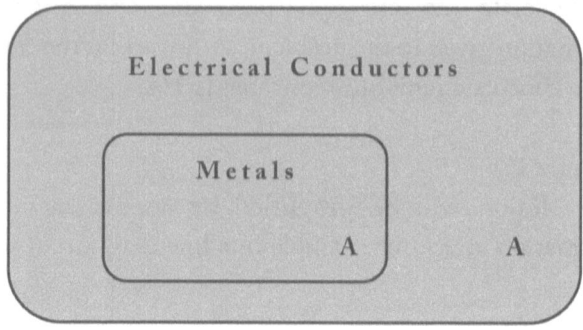

The diagram shows that A can be an electrical conductor without it being a metal. Hence it would not be logical to conclude that A has to be a metal.

Exercise 2.8

Statement "Some animals are warm-blooded" can be represented by the diagram (a). Diagram (b) includes the information in the second statement "A is not warm-blooded" in diagram (a). Diagram (b) shows that A can be an animal (though it need not be an animal) and hence it cannot be concluded that A is not an animal

Thinking Strategies for Solving Problems

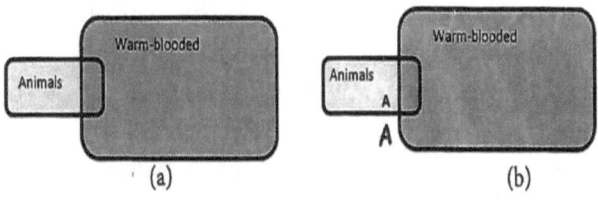

(a)　　　　　　　　　　(b)

Exercise 2.9

The diagram below, which shows the positions of all trains travelling from B towards A at 8:00 am, will assist the solution.

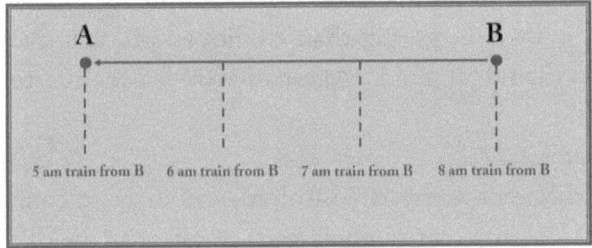

The 5 am train from B would have just reached A at 8 am (since travel from B to A needs 3 hours) and the 6 am and 7 am trains will respectively be 50 km and 100 km from A (because speed of train is 50 km per hour). These three trains will meet the train that leaves A at 8 am which, in addition, will meet the trains that leave B at 8 am, 9 am, 10 am and 11 am. Hence 7 trains will be met by the train that leaves A at 8 am.

The times at which the 8 am train from A will meet the 7 trains from B can be easily calculated and are:

8 am (for the 5 am train); 8.30 am (for 6 am train); 9 am (for 7 am train); 9.30 am (for 8 am train); 10 am (for 9 am train); 10.30 am (for10 am train); 11 am (for 11 am train).

Exercise 2.10
To help avoid errors, use the defining equation for c for making the required deduction. The equation should be written

$$c_{alcohol} = n_{alcohol} / V_{solution} \quad \text{(and not as } c = n/V\text{)}$$

Since $V_{solution}$ is greater than V_{water} (because $V_{solution} = V_{water} + V_{alcohol}$), it will be greater than 2.0 litres and c will therefore be less than 0.50 mol l^{-1}. Alternative (b) is hence correct.

Exercise 2.11
Four different masses are involved, and to avoid confusion and errors different symbols must be given to the different masses:

$$m_A(\text{at } 20°C) \;;\; m_A(\text{at } 300°C) \;;\; m_B(\text{at } 300°C) \;;\; m_C(\text{at } 300°C)$$

These four masses are related to each other by the law of conservation of mass. Since total mass before reaction (at 20°C) is equal to the total mass after reaction (at 300°C) we may write

$$m_A(\text{at } 20°C) = m_A(\text{at } 300°C) + m_B(\text{at } 300°C) + m_C(\text{at } 300°C)$$
$$m_A(\text{at } 300°C) = m_A(\text{at } 20°C) - m_B(\text{at } 300°C) - m_C(\text{at } 300°C)$$
$$= 5g - 1g - 0.8g = 3.2g$$

Exercise 2.12

In an aqueous solution of HCl at 25°C, c_H^+ can never be less than 1×10^{-7} M because H⁺ ions are also formed by dissociation of water, $H_2O \rightarrow H+(aq) + OH^-(aq)$.

The concentration of H⁺ ions produced by the dissociation of water, at 25°C, is 1×10^{-7} M.

CHAPTER 3
IDENTIFICATION AND FOCUSING ON THE GOAL

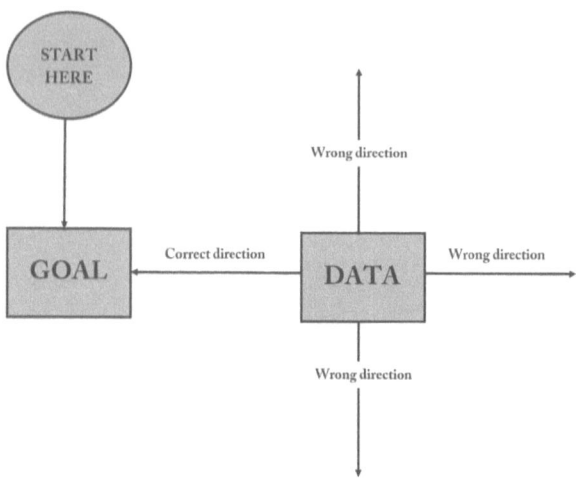

Start solutions of problems from the goal. If we start with the data, there is a greater chance of going in the wrong direction than in the correct direction.

Thinking Strategies for Solving Problems

A crucial step for successful problem solving is identification of the goal and then focusing sharply on it. One should generally start the solution of a problem from the goal and work "backwards" towards the data: the data should be checked only to see whether the quantities needed to reach the goal are given there (Fig 3.1).

Difficulties with problem solving are often due to our not identifying the goal and starting the solution by asking the question: what needs to be done to reach the goal? If, for example, the goal is the calculation of some physical quantity, it is logical that the solution should start with the definition (if possible, the defining equation) of that quantity. Many students, however, do not do this. Instead, they attempt to manipulate either the data given or the methods they remember

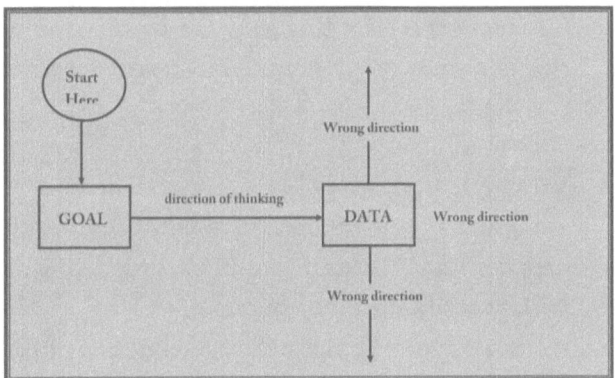

Fig 3.1: Where to start the solution of a problem
Start from the goal: for example with the definition of the required quantity. The figure shows that if we start with the data there is a greater chance of going in the "wrong direction" than in the "correct direction".

Mailoo Selvaratnam

The importance of focusing sharply on the goal for effective problem solving will be illustrated using some examples that have been tested on many groups of first year university students. You should first try to solve the examples on your own before reading through the solutions provided. This will help you to obtain a greater appreciation of the importance of this strategy. To encourage you to do this, the questions in the examples are given below.

3.1. Two men A and B, who are separated by a distance of 7.5 km, walk towards each other at speeds of 3.0 km per hour and 2.0 km per hour. A dog also starts with man A and runs at a constant speed of 6.0 km per hour to meet B, then runs back to A and repeats the process of running between A and B. Calculate the total distance run by the dog when the men meet in 1.5 hours. (Note: speed = distance travelled/time).

3.2. 5.0 grams of gaseous N_2O_4 are present in a 2 litre vessel at 20°C. When heated to 300°C, with volume kept constant, 2.0 grams of N_2O_4 breaks down into NO_2 according to the equation N_2O_4 (g) → $2NO_2$ (g). Calculate the density of the gaseous mixture present in the vessel at 300°C. (Density = mass/volume; total mass does not change during any chemical reaction).

3.3. The density of a gas present in a closed vessel at a temperature T is x. If the temperature is doubled to $2T$, with volume kept constant, which one of the following will be its density?

(a) $0.5x$ (b) x (c) $2x$ (d) $4x$ (e) x^2

3.4. State clearly how you will prepare 1.00 dm^3 of an HCl solution having the concentration 0.250 mol dm^{-3}, starting with a 0.500 mol dm^{-3} HCl solution ($c = n/V$)

Mailoo Selvaratnam

EXAMPLE 3.1

Two men A and B, who are separated from one another, walk towards each other at speeds of 3.0 km per hour and 2.0 km per hour. A dog also starts with man A and runs at a constant speed of 6.0 km per hour to meet B, then runs back to A and repeats the process of running between A and B. Calculate the total distance run by the dog when the men meet in 1.5 hours. (Note: speed = distance travelled/ time).

Solution

This problem is very easy to solve if its solution is started with the defining equation for the required quantity, which is $d = s\,t$. This equation shows that the distance d travelled by the dog depends *only* on the speed (s) of the dog and the time (t) for which the dog runs. Hence

$$d = s\,t$$
$$= (6.0 \text{ km h}^{-1})\, 1.5 \text{ h} = 9.0 \text{ km}$$

This question, which is a type of question often used in IQ tests, was used for testing first year university students at some workshops.

Many students had difficulty with the problem because they were *distracted by the irrelevant information* (speeds of the men and distance separating them) which they tried to use for solving the problem.

Thinking Strategies for Solving Problems

EXAMPLE 3.2

5.0 grams of gaseous N_2O_4 is present in a 2 litre vessel at $20°C$. When heated to $300°C$, with volume kept constant, 2 grams of N_2O_4 breaks down into NO_2 according to the equation N_2O_4 (g) $\rightarrow$ $2NO_2$ (g). Calculate the density of the gaseous mixture present in the vessel at $300°C$. (Density = mass/volume; total mass does not change during any reaction).

Solution

This problem is easy to solve if its solution is started with the defining equation for the required quantity, $d = m/V$.

This equation shows that the calculation of density needs only m (mass) and V (volume) and since these are given in the data (m = 5.0 g, V = 2.0 l), d is easily calculated to be 2.5 g l^{-1}

This problem (and similar problems) has been tested with many groups of first year university students and it was found that many of them could not solve it correctly. The major reason for their difficulty was associated with their trying to manipulate the irrelevant data (equation for the dissociation of N_2O_4; mass of N_2O_4 dissociated).

This type of difficulty cannot, however, occur if the solution to a problem is started with the defining equation for the required quantity.

EXAMPLE 3.3

The density of a gas present in a closed vessel at a temperature T is x. If the temperature is doubled to $2T$, with volume kept constant, which one of the following will be its density?

(a) $0.5x$ (b) x (c) $2x$ (d) $4x$ (e) x^2

Solution

This problem too is easy to solve if we start its solution by focusing on the defining equation for density, $d = m/V$, which shows that density depends only on its mass m and volume V.

In this problem m and V are constant and therefore d will not change from its initial value of x.

More than a half of first year university students tested with this problem thought, however, that density of the gas will change (either double or halve) when its temperature is doubled. Their error may be attributed to their not using the defining equation for density for deducing the answer.

> *Many students assume illogically that if some quantity (e.g. temperature) is doubled, another quantity should also be doubled or halved.*

Thinking Strategies for Solving Problems

EXAMPLE 3.4

State clearly how you will prepare 1.00 dm³ of an HCl solution having the concentration 0.250 mol dm⁻³ starting with a 0.500 mol dm⁻³ HCl solution ($c = n / V$).

Solution

It is best to start the solution by focusing on the goal which is the preparation of 1.00 dm³ of a 0.250 mol dm⁻³ HCl solution. This solution contains 0.250 mole of HCl (apply equation $n = c \times V$ = 0.250 mol dm⁻³ × 1.00 dm³ = 0.250 mol).

0.250 mole of HCl must therefore be withdrawn from the given 0.500 mol dm⁻³ solution, and this corresponds to 0.500 dm³ of the solution (apply again the equation $n = c\,V$ from which $V = n / c$ = 0.250 mol / 0.500 mol dm⁻³ = 0.500 dm³ = 500 cm³).

500 cm³ of the given 0.500 mol dm⁻³ HCl solution must therefore be diluted with water to 1.00 dm³ (i.e. 1000 cm³) to prepare the required solution.

Comment

The solution above, which was given in a step-by-step manner, can be simplified because the amount (n) of HCl always remains constant when any solution is diluted. Application of $c = n / V$ to the initial solution (*i.s*) and final solution (*f.s*) will then show that c (*i.s*) V (*i.s*) = c (*f.s*) V (*f.s*), from which V (*i.s*) = c (*f.s*) V (*f.s*) / c (*i.s*) = 500 cm³.

EXERCISES

Density (d) = mass/volume; concentration (c) = amount (moles)/volume; an ideal gas obeys the equation $pV = nRT$; speed = distance travelled/time; M = concentration in mol dm^{-3}.

Exercise 3.1
Car A leaves Mmabatho for Johannesburg at an average speed of 90 kilometers per hour. An hour later, another car B leaves Johannesburg for Mmabatho at an average speed of 120 kilometers per hour. The distance between Mmabatho and Johannesburg is 300 kilometers. When the cars meet, which car (A or B) will be closer to Mmabatho?

Exercise 3.2
There are 12 coins in a dozen one rand coins. How many coins are there in a dozen five rand coins?

Exercise 3.3
A box contains 30 blue socks, 20 black socks and 10 brown socks. How many socks have to be taken out to have a pair that has (a) the same colour (b) brown colour

Exercise 3.4
A car travels from A to B at an average speed of 90 km h^{-1} (kilometres per hour). The distance between A and B is

180 km. For the return journey from B to A, average speed is 60 km h^{-1}. Calculate the average speed of the car for the entire journey

Exercise 3.5
The density of 1g of a solid is x. Which one of the following will be the density of 2 g of this solid?   (a) $0.5x$ (b) x (c) $2x$ (d) x^2

Exercise 3.6
Will the density of a gas increase, decrease or remain unchanged if its temperature is increased when (*a*) the volume of the gas is kept constant (*b*) the pressure of the gas is kept constant?

Exercise 3.7
A closed 2.0 litre vessel contains 5.0 grams of a gas A at 25° C. When the gas is heated to 200° C, 1.5 grams of it breaks down into two other gases B and C according to the equation A (g) → B (g) + C (g). Calculate the density of the gas mixture at 200° C.

Exercise 3.8
A closed vessel has a mixture of two gases A and B at a temperature T and pressure p. The gases do not react with each other. The mole fraction (x_A) of A in the gaseous mixture is 0.20. What will be the mole fraction of A in the gas mixture if: (a) its temperature is doubled from T to $2T$; (b) its pressure is doubled from p to $2p$. ($x_A = n_A / n_T$, where n_A and n_T are respectively the moles of A and the total moles of A and B).

Exercise 3.9
An aqueous solution of sugar at $20°C$ is warmed to $30°C$. Will the concentration of sugar in the solution then increase, decrease or remain unchanged?

Exercise 3.10
Will the concentration of an ideal gas increase, decrease or remain unchanged if its temperature is increased when: (a) the volume of the gas is kept constant; (b) the pressure of the gas is kept constant?

Exercise 3.11
You are given a 1.25 mol dm^{-3} HCl solution. State clearly how you will prepare, from this solution, 250 cm^3 of a 0.100 mol dm^{-3} HCl solution.

Exercise 3.12
Consider 25 cm^3 of a 1.0 M HCl solution. Will the concentration of HCl increase, decrease or remain unchanged if to it is added:
(a) 5 cm^3 of water;
(b) 5 cm^3 of 1.0 M HCl solution;
(c) 5 cm^3 of 1.0 M CuSO$_4$ solution;
(d) 5 cm^3 of 1.0 M NaOH solution.

Thinking Strategies for Solving Problems

SOLUTIONS TO EXERCISES

Exercise 3.1
Focus on the goal. When the cars meet, both cars will be the same distance away from Mmabatho.

Exercise 3.2
Focus on the word "dozen". Answer is 12 coins

Exercise 3.3
Principle to be used: consider the *worst possibility*.

(a) 3 socks taken out need not give a pair of the same colour since their colours could be blue, black and brown. Hence 4 socks must be taken out.

(b) 51 socks taken out could be 30 blue socks, 20 black socks and 1 brown sock. Hence 52 socks must be taken out to have a pair of brown socks.

Exercise 3.4
The average speed of the car for the entire journey is not equal to the average value of 90 km h^{-1} and 60 km h^{-1}.

Focus on the goal: on the average speed (*s*) for which the defining equation is $s = d/t$, where d = total distance travelled and t = total time.

$$s = \frac{d}{t} = \frac{180 \text{ km} + 180 \text{ km}}{2\text{h} + 3\text{h}} = 72 \text{ km h}^{-1}$$

Exercise 3.5
Density of 2g of the solid will be the same as that of 1g of the solid. This is because when mass (*m*) is doubled, volume (*V*) too will be doubled and hence *m/V*, which is density, will not change.

Exercise 3.6
An important strategy that should always be used during problem solving is to *focus sharply on the definition of the required quantity*. Equation $d = m/V$ shows that

(a) *d* depends *only* on *m* and *V*. Since *m* and *V* do not change in this problem, *d* will not change.

(b) *d* will decrease. This is because, when *p* is kept constant, an increase in temperature will increase *V*.

Exercise 3.7
Note that $m_{mixture}$ is equal to the initial mass of A (by the law of conservation of mass). Hence the density of the gas mixture will be 5 g / 2 l = 2.5 g l^{-1}

Thinking Strategies for Solving Problems

Exercise 3.8
Focus on the defining equation for mole fraction of A, $x_A = n_A / n_T$, which shows that x_A depends only on n_A and n_T and does not depend on p or T. x_A will therefore not change when T and p are changed.

Exercise 3.9
Focus on the defining equation $c_{sugar} = n_{sugar} / V_{solution}$.

When T is increased, V will increase and therefore c will decrease.

Exercise 3.10
Focus on the defining equation $c = n / V$ to answer the question.

(a) This equation shows that if V is kept constant and since n also does not change, c will not change.

(b) To deduce how c will change when T is changed, at constant p, it is necessary to derive an equation that shows the relationship between c, p and T. This can be done by replacing the V term in $c = n/V$ by nRT/p (rearrangement of $pV = nRT$ gives $V = nRT/p$)

$$c = \frac{n}{V} = \frac{n}{nRT/p} = \frac{p}{RT}$$

The equation $c = p / RT$ shows that if T is increased c will decrease when p is kept constant (R is always a constant).

Exercise 3.11
Start the solution by focusing on what needs to be done: the preparation of 250 cm³ of 0.100 mol dm⁻³ HCl solution. This solution contains 0.0250 mol HCl (apply equation $n = c \times V = 0.100$ mol dm⁻³ × 250 × 10⁻³ dm³ = 0.0250 mol).

0.0250 mol HCl must therefore be withdrawn from the given 1.25 mol dm⁻³ solution and this corresponds to 0.0200 dm³ HCl solution ($V = n/c = 0.0250$ mol/1.25 mol dm⁻³ = 0.0200 dm³ = 20.0 cm³)

20.0 cm³ of the given 1.25 mol dm⁻³ HCl solution must therefore be diluted with water to 250 cm³ to prepare 250 cm³ of 0.100 mol dm⁻³ HCl solution.

Exercise 3.12
Focus on the defining equation for the concentration of HCl, written explicitly as $c_{HCl} = n_{HCl}/V$. This equation shows that c_{HCl} can change only if n_{HCl} or $V_{solution}$ changes. Therefore

(a) c_{HCl} will decrease because $V_{solution}$ is increased;
(b) c_{HCl} will not change because the same solution is added;
(c) c_{HCl} will decrease because $V_{solution}$ is increased;
(d) c_{HCl} will decrease because n_{HCl} will decrease and $V_{solution}$ is increased.

CHAPTER 4

IDENTIFICATION OF PRINCIPLES NEEDED FOR SOLUTIONS

| 2.00 GRAMS OF LIQUID A | + | 2.00 GRAMS OF LIQUID B | = | 4.00 GRAMS |

| 2.00 LITRES OF LIQUID A | + | 2.00 LITRES OF LIQUID B | = | ? |

Was your answer 4.00 litres?

This is not correct.

Why?

Mailoo Selvaratnam

Calculations and deductions should always be based on explicitly identified principles and laws. This strategy will guide, simplify and sharpen problem solving and also help avoid errors. It will also help students obtain greater insight and understanding of subject-content knowledge, which is an important objective of problem solving in academic courses.

The importance of explicitly identifying and using principles for problem solving will be illustrated with some examples that have been tested on many groups of first year university students. The questions in these examples are given below to encourage you to try to solve them on your own, before reading through the solutions.

4.1. You are given a 9 litre vessel, a 4 litre vessel, an empty vessel A and water. State how you will measure 6 litres of water into the vessel A (Note: Using the 9 litre vessel and 4 litre vessel you can measure only 9 litres and 4 litres of water respectively).

4.2. The human population in a country was 46 million in January 2009 and 50 million in January 2011. Which one of the following would you expect to be the population, assuming same rate of population growth, in January 2013?

(a) 54 million (b) less than 54 million
(c) greater than 54 million.

Thinking Strategies for Solving Problems

4.3. 16 sticks of the same size, numbered 1-16, are arranged in the diagram below to form five squares. Rearrange any three of these sticks to obtain four same-sized squares. State which sticks you will remove, and show how you will rearrange them.

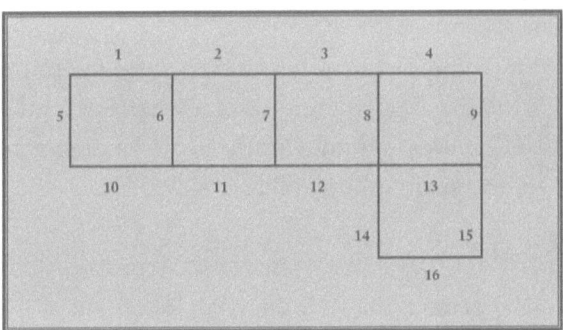

4.4. The figure below shows an arrangement of 9 dots. Show how all these dots can be connected together by drawing 4 straight lines without lifting the pencil.

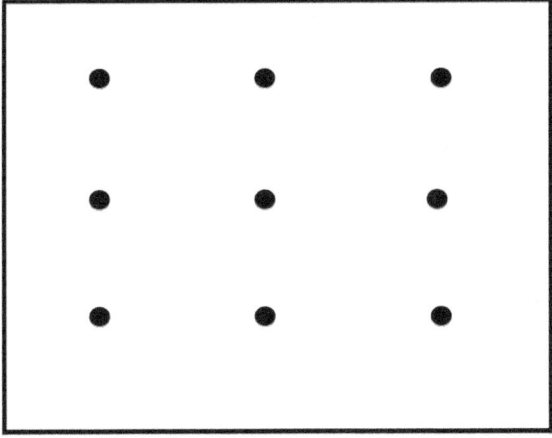

4.5. A closed vessel contains a mixture of two gases A and B at a temperature T and pressure p. The gases do not react with each other. The mass of A is 1.2 grams and the total mass of A and B is 2.0 grams. What will be the mass of A if: (a) the pressure is doubled to $2p$; (b) the temperature is doubled to $2T$?

4.6. The volume of a liquid increases when its temperature is increased. If the volume of a sample of liquid water is 10 cm³ at a temperature T, its volume when the temperature is doubled to $2T$ will be:

(a) 5 cm³ (b) 10 cm³ (c) 20 cm³ (d) greater than 10 cm³ but less than 20 cm³ (e) slightly less than 10 cm³

4.7. Hydrogen peroxide (H_2O_2) decomposes according to the equation $2\,H_2O_2 \rightarrow 2\,H_2O\,(l) + O_2\,(g)$, with the rate of decomposition being directly proportional to the concentration of H_2O_2.

(a) If the volume of O_2 evolved, when a H_2O_2 solution decomposes, is 50 cm³ after one hour, which one of the following will be the volume of O_2 evolved after two hours?

(i) 25 cm³
(ii) 50 cm³
(iii) 100 cm³
(iv) in-between 50 cm³ and 100 cm³
(v) greater than 100 cm³

(b) If the initial concentration of H_2O_2 is 1.0 M and its concentration is 0.8 M after 1 hour, which one of the following will be its concentration after 2 hours?
(i) 0.6 M
(ii) less than 0.6 M
(iii) greater than 0.6 M

EXAMPLE 4.1

You are given a 9 litre vessel, a 4 litre vessel, an empty vessel A and water. State how you will measure 6 litres of water into the vessel A. (Using the 9 litre vessel and 4 litre vessel you can measure only 9 litres and 4 litres of water respectively).

Solution

A *principle* that can be used to simplify the solution of this problem is that the difference between multiples of 9 ℓ (ℓ = litre) and multiples of 4 ℓ must be equal to 6 ℓ. This is because from the 9 ℓ and 4 ℓ vessels we can measure only 9 ℓ, 4 ℓ and multiples of 9 ℓ and 4 ℓ. The table below gives some multiples of 9 ℓ and 4 ℓ.

Multiples of 9 ℓ	Multiples of 4 ℓ
1 x 9 ℓ = 9 ℓ	1 x 4 ℓ = 4 ℓ
2 x 9 ℓ = 18 ℓ	2 x 4 ℓ = 8 ℓ
3 x 9 ℓ = 27 ℓ	3 x 4 ℓ = 12 ℓ
4 x 9 ℓ = 36 ℓ	4 x 4 ℓ = 16 ℓ
5 x 9 ℓ = 45 ℓ	5 x 4 ℓ = 20 ℓ
6 x 9 ℓ = 54 ℓ	6 x 4 ℓ = 24 ℓ

From the multiples given above, 6 ℓ can be obtained in two ways.

(a) 18 ℓ - 12 ℓ (i.e. 2 x 9ℓ - 3 x 4 ℓ)

(b) 24 ℓ - 18 ℓ (i.e. 6 x 4 ℓ - 2 x 9 ℓ)

Thinking Strategies for Solving Problems

Method (a) is simpler. Practically, we can introduce 18 ℓ of water into vessel A (by adding 9 ℓ twice) and then remove 12 ℓ from it by pouring out thrice into the 4 ℓ vessel.

Most people solve this problem using a *trial and error approach*. This takes more time and is more difficult. Also the "learning" associated with the solution does not often carry-over to improve the solution of similar types of problems.

EXAMPLE 4.2

The human population in a country was 46 million in January 2009 and 50 million in January 2011. Which one of the following would you expect to be the population, assuming same rate of population growth, in January 2013?

(a) 54 million (b) less than 54 million (c) greater than 54 million.

Solution

Alternative (c) is correct, not alternative (a).

The *principle* that has to be used is that increase in population is directly proportional to the population.

Since the population progressively increases, the rate at which population increases should also progressively increase.

For example, in January 2009 the increase in population is directly proportional to 46 million while in January 2011 the increase in population will be larger because it is directly proportional to a larger number (50 million).

Thinking Strategies for Solving Problems

EXAMPLE 4.3

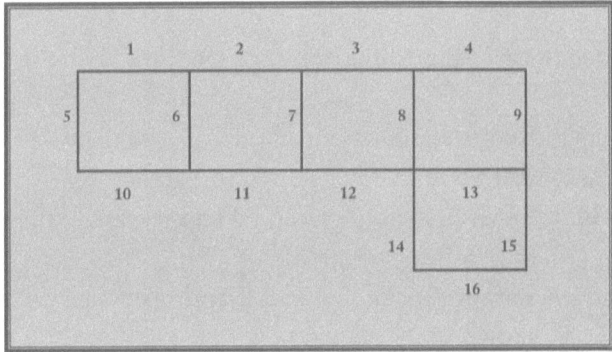

16 sticks of the same size, numbered 1-16, are arranged in the figure above to form five squares. Rearrange any three of these sticks to obtain four same-sized squares. State which sticks you will remove, and show how you will rearrange them.

Solution

Most people try to solve this problem by *a trial and error method*. A much better method would be to identify *a principle* that can be used to obtain the solutions. Since a square needs 4 sticks, it should be clear that 4 squares can be formed from 16 sticks only if no stick is a common side to two squares.

In the figure shown, sticks 6, 7, 8 and 13 are the only ones that belong to two squares (stick 6, for example, is common to the two squares on the left side of the figure). The problem therefore is to remove three sticks so that

Mailoo Selvaratnam

sticks 6, 7, 8 and 13 do not belong to two squares, and then rearrange these sticks to obtain four squares.

Four solutions are possible and they are:

(a) Remove sticks numbered 2, 11 and 4 and rearrange them as shown in Fig. (a)
(b) Remove sticks 2, 11 and 9 and rearrange them as shown in Fig. (b)
(c) Remove sticks 2, 4 and 9 and rearrange them as shown in Fig. (c)
(d) Rearrange sticks 4, 11 and 9 and rearrange them as shown in Fig.(d)

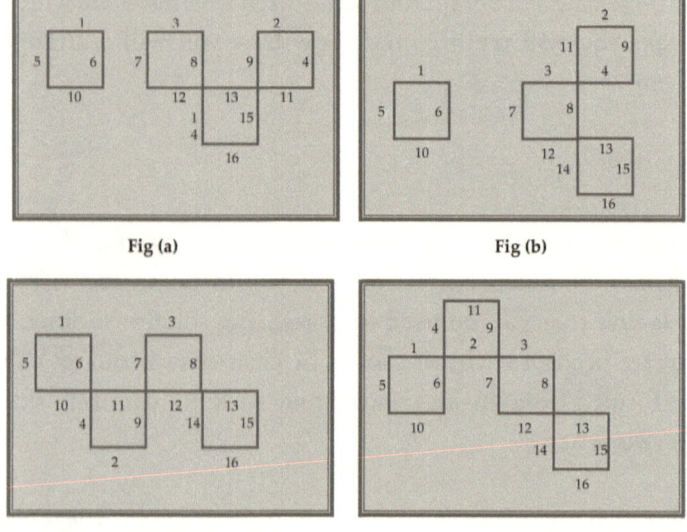

Fig (a) Fig (b) Fig (c) Fig (d)

Thinking Strategies for Solving Problems

EXAMPLE 4.4

The figure below shows an arrangement of 9 dots. Show how all these dots can be connected together by drawing 4 straight lines without lifting the pencil.

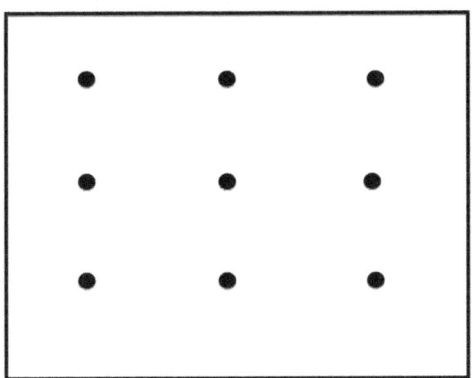

Solution

Two *principles* need to be used to solve this problem:

(a) The lines should be extended to go outside the "box" formed by the 9 points;

(b) For 4 lines to be drawn (without lifting the pencil) through the 9 points, one line must pass through 3 points and each of the other three lines must pass through 2 new points (since 9 = 3+ 2+2+2).

In the diagrams below, the first line connects 3 dots, the second line connects 2 new dots, the 3rd line 2 new dots and the 4th line 2 new dots.

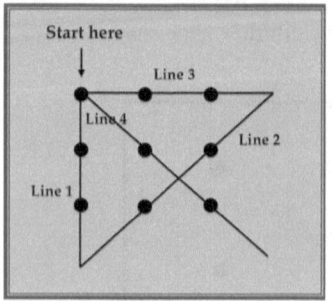 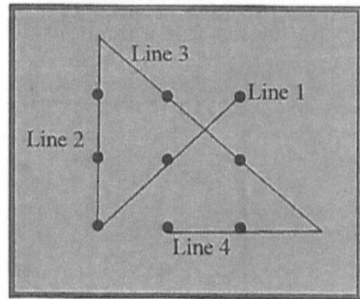

Thinking Strategies for Solving Problems

EXAMPLE 4.5

A closed vessel contains a mixture of two gases A and B at a temperature T and pressure p. The gases do not react with each other. The mass of A is 1.2 grams and the total mass of A and B is 2.0 grams. What will be the mass of A if: (a) the pressure is doubled to $2p$; (b) the temperature is doubled to $2T$?

Solution

Mass is a basic physical quantity and hence will not depend on any other quantity. Mass will therefore not change if pressure or temperature is changed.

More than a half of first year university students tested, however, thought incorrectly that mass will change (either double or halve) when pressure or temperature is doubled. These students evidently did not recognize the importance of the following principle:

> *To deduce how some quantity changes when another quantity is changed, it is always necessary to make use of a principle that relates the two quantities.*

If no principle exists, or if the principle is not identified, valid deductions cannot be made.

A common mistake often made by students is to assume, without any reason, that two quantities have to be either directly proportional or inversely proportional to one another.

Mailoo Selvaratnam

EXAMPLE 4.6

The volume of a liquid increases if its temperature is increased. If the volume of a sample of liquid water is 10 cm^3 at a temperature T, its volume when the temperature is doubled to $2T$ will be:

(a) 5 cm^3 (b) 10 cm^3 (c) 20 cm^3 (d) greater than 10 cm^3 but less than 20 cm^3 (e) slightly less than 10 cm^3.

Solution

Though the volume of a liquid increases when its temperature is increased, it is not directly proportional to temperature. From experience we know that volume of water increases only slightly when temperature is increased.

Hence the correct answer is alternative (d).

About a third of the students tested, however, thought incorrectly that volume will be doubled when temperature is doubled. They implicitly assumed, without any justification, that liquid volume is directly proportional to temperature.

Thinking Strategies for Solving Problems

EXAMPLE 4.7

Hydrogen peroxide (H_2O_2) decomposes according to the equation 2 H_2O_2 (l) → 2 H_2O (l) + O_2 (g), with the rate of decomposition being directly proportional to the concentration of H_2O_2.

(a) If the volume of O_2 evolved when a H_2O_2 solution decomposes is 50 cm³ after one hour, which one of the following will be the volume of O_2 evolved after two hours?

 (i) 25 cm³
 (ii) 50 cm³
 (iii) 100 cm³
 (iv) in-between 50 cm³ and 100 cm³
 (v) greater than 100 cm³.

(b) If the initial concentration of H_2O_2 is 1.0 M and its concentration is 0.8 M after 1 hour, which one of the following will be its concentration after 2 hours?
 (i) 0.6 M
 (ii) less than 0.6 M
 (iii) greater than 0.6 M ?

Mailoo Selvaratnam

Solution

(a) Correct alternative is (iv).

Many students select alternative (iii) as correct. They incorrectly assumed that volume of oxygen evolved is directly proportional to decomposition time: that if decomposition time is doubled from 1 hour to 2 hours, the volume of oxygen evolved will double from 50 cm³ to 100 cm³. Volume of oxygen evolved, however, is not directly proportional to decomposition time.

The *principle* that must be used for making the deduction is that the volume of oxygen evolved at any time is directly proportional to the concentration of H_2O_2.

As decomposition proceeds, the concentration of H_2O_2 decreases and hence the rate at which O_2 is evolved will progressively decrease with time. Hence volume of O_2 evolved will not double to 100 cm³ but will be less than 100 cm³.

(b) Correct alternative is (iii).

This is because the *rate* at which concentration of H_2O_2 decreases will progressively decrease with time because the concentration of H_2O_2 progressively decreases with time.

Thinking Strategies for Solving Problems

EXERCISES

Mole fraction of A = moles of A / total moles

Exercise 4.1
128 players take part in a singles knockout tennis tournament. Find the total number of matches that have to be played to find the champion (winner).

Exercise 4.2
You are given 9 litre and 4 litre vessels, another vessel A and water.
State how you will measure into A the following volumes of water
(a) 1 litre (b) 2 litres (c) 3 litres (d) 7 litres

Exercise 4.3
20 sticks of the same size, labelled 1- 20, are arranged in the figure below to form 7 squares. Rearrange any three of these sticks so that five same- sized squares are obtained. Show all the possible ways of rearranging the sticks.

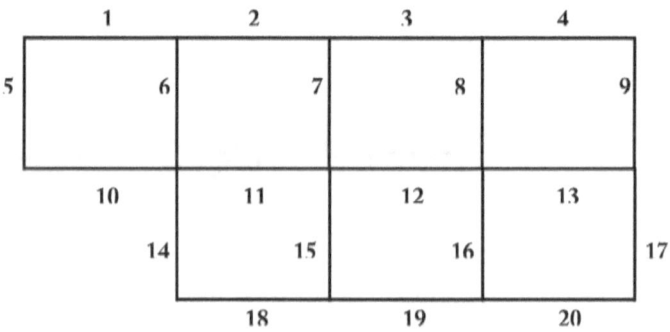

Exercise 4.4

Connect the 16 dots in the figure below by drawing 6 lines without lifting your pencil.

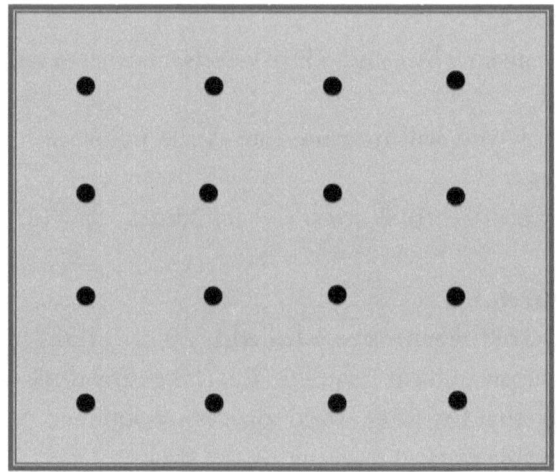

Exercise 4.5

Show how 12 trees can be planted in 6 straight lines, with each line having 4 trees.

Thinking Strategies for Solving Problems

Exercise 4.6
One hundred rands are invested in a bank at the interest rate of 10%, the interest being calculated annually and added to the capital. Calculate the total value of the interest after:

(a) one year, (b) two years, (c) three years.

Exercise 4.7
A person tosses a coin and gets "Head". He tosses the coin again and he then also gets Head. If he tosses the coin a third time, will there be a greater chance or probability of getting Tail than Head?

Exercise 4.8
When a stone is dropped from a height, it travels 32 feet after 1 second. Which one of the following will be the distance travelled after 2 seconds?

(a) 64 feet, (b) less than 64 feet, (c) greater than 64 feet

Exercise 4.9
A compound is present in a closed vessel. If 0.2 g of the compound decomposes after 1 hour, which one of the following will be the mass that decomposes after 2 hours?

(a) 0.4 g, (b) less than 0.4 g, (c) greater than 0.4 g

Exercise 4.10
When 2.000 cm³ of a liquid A are mixed with 2.000 cm³ of another liquid B, which one of the following will be the final volume?
(a) 4.000 cm³
(b) less than 4.000 cm³
(c) greater than 4.000 cm³
(d) volume cannot be calculated accurately using the data given.
State the reasoning you used to obtain the answer.

Exercise 4.11
The mass fraction of a gas A in a mixture of gases at 25°C and 1.0 atm is 0.20. What will be its mass fraction if: (a) the pressure is doubled to 2 atm; (b) the temperature is increased to 50°C; (c) the volume of the vessel is doubled.

Exercise 4.12
Person A needs 6 days to paint the walls of a house and B needs 12 days. How many days will be needed if both A and B jointly paint the house?

Thinking Strategies for Solving Problems

SOLUTIONS TO EXERCISES

Exercise 4.1
The use of the following *principle* will make the solution very easy. Since all the players except the champion have to lose a match, the total number of losers, and hence the total number of matches played, must be equal to the total number of participants minus one. Since the number of participants is 128, the total number of matches played will be 128 - 1 = 127.

This problem is not difficult to solve, but almost all the people tested (at workshops) solved it by adding together the numbers of first-round matches (64), second-round matches (32), third-round matches (16), fourth-round matches (8), quarter-final matches (4), semi-final matches (2) and the final match (1). This takes much more time and is cumbersome, and sometimes leads to careless errors.

Exercise 4.2
A *principle* that could be used to solve this problem is that from 9 ℓ and 4 ℓ vessels we can measure only 9 ℓ, 4 ℓ and multiples of 9 ℓ and 4 ℓ.

Multiples of 9	Multiples of 4
1 x 9 = 9	1 x 4 = 4
2 x 9 = 18	2 x 4 = 8
3 x 9 = 27	3 x 4 = 12
4 x 9 = 36	4 x 4 = 16
5 x 9 = 45	5 x 4 = 20
6 x 9 = 54	6 x 4 = 24
	7 x 4 = 28

From the multiples given above, the possible ways of obtaining the required volumes (1 ℓ, 2 ℓ, 3 ℓ, 7 ℓ) are

(a)   1 = 9 - 8 (i.e. 9 - 4 - 4)
       1 = 28 - 27 (i.e. 7 x 4 - 3 x 9)

(b)   2 = 18 - 16 (i.e. 2 x 9 - 4 x 4)
       2 = 20 - 18 (i.e. 5 x 4 - 2 x 9)

(c)   3 = 12 - 9 (i.e. 3 x 4 - 9)
       3 = 27 - 24 (i.e. 3 x 9 - 6 x 4)

(d)   7 = 16 - 9 (i.e. 4 x 4 - 9)
       7 = 27 - 20 (i.e. 3 x 9 - 5 x 4)

Thinking Strategies for Solving Problems

Exercise 4.3

Use the following *principle* to solve the problem: since each square needs 4 sticks, to form 5 squares from 20 sticks no stick must be a common side to two squares.

To ensure that no stick is a common side in two squares, either sticks 2, 19 and 4 or 2, 19 and 9 must be removed. They could be arranged as shown below to obtain 5 same-sized squares.

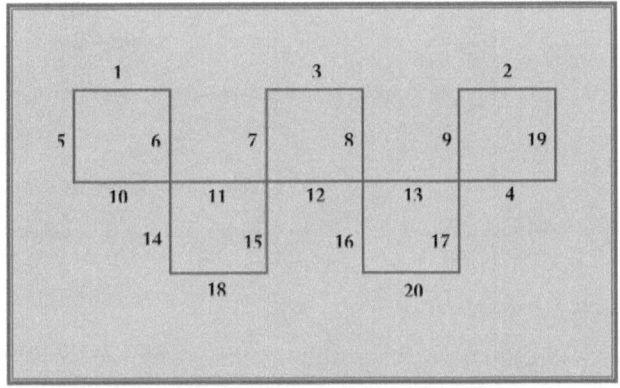

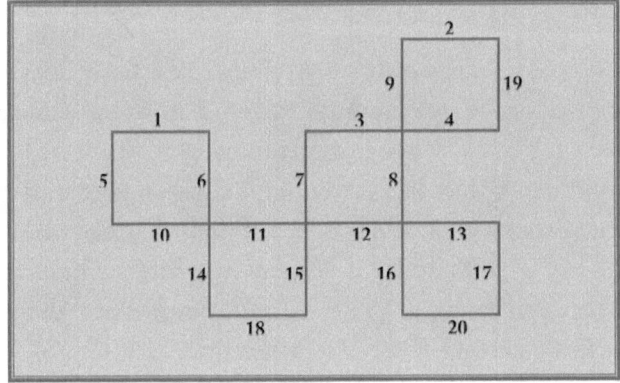

Exercise 4.4
See Example 4.4 to identify principles used (16 dots = 4+3+3+2+2+2 "new" dots per line).

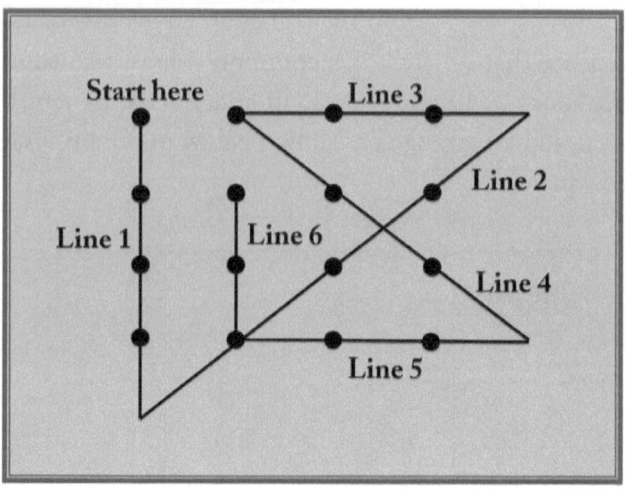

Exercise 4.5
The arrangement of the six lines connecting the trees must be symmetrical (intuitive judgment).

If just one line passes through each tree, the 6 lines will pass through 24 trees because each line must pass through 4 trees (fig. a, each tree is represented by •). To have a *half* the number of trees (i.e. 12 trees) in 6 lines, it is necessary to double (from 1 to 2) the number of lines that pass through each tree. That is, to have 12 trees in 6 lines, 2 lines must pass through each tree (6 x 2 = 12). A symmetrical diagram of 6 lines joining 6 trees is a hexagon (see fig. b). The extension of these 6 lines till they meet (fig. c) will give the

Thinking Strategies for Solving Problems

required arrangement of 6 lines passing through 12 trees, each line having 4 trees.

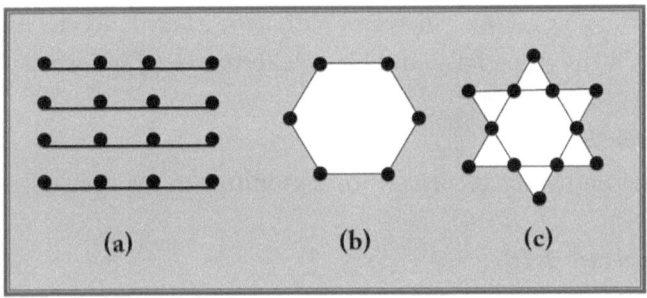

Exercise 4.6
Principle: Interest depends not only on time but also on the capital. Capital increases each year and hence interest earned each year will increase from year to year.

- (a) Interest will be 10 rands (10% of 100 rands)
- (b) Total interest will be 21 rands (10% of 100 rands + 10% of 110 rands)
- (c) Interest will be 33.1 rands (10% of 100 rands + 10% of 110 rands + 10% of 121 rands)

Exercise 4.7
No. The chance of getting "Head" will be the same as the chance of getting "Tail". The coin cannot evidently "remember" what happened to it previously.

Principle: The probability of a particular outcome in a random event does not depend on previous events.

Exercise 4.8
Alternative (a) is not correct because distance (d) travelled, $d = s\,t$, is not directly proportional to time (t) because speed (s) is not constant. s increases with time (there is acceleration when the stone falls) and hence alternative (c) is correct.

Exercise 4.9
Alternative (b) is correct (for reasoning see example 4.7).

Exercise 4.10
Correct answer is alternative (d).

Volume, unlike mass and charge, is not an additive property and therefore it cannot be concluded that when two liquids are mixed together the final volume will be equal to the sum of the volumes of the two liquids. An example to illustrate why volumes cannot be added together is shown visually below. If the molecules in liquid A are much larger than the molecules in liquid B, the small molecules in liquid B can occupy the empty spaces (see fig) between the large molecules in liquid A, and therefore the volume of the solution will be less than the sum of the volumes of liquids A and B.

The smaller B molecules can occupy the empty spaces between the larger A molecules.

Thinking Strategies for Solving Problems

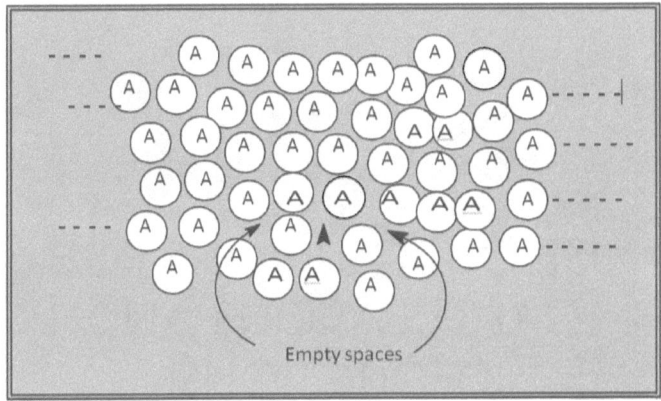

Exercise 4.11

Focus on the defining equation for the mass fraction of A (which is m_A / m_{total}), which shows that it depends only on m_A and total mass. Since masses do not depend on pressure, temperature or volume, mass fraction will not depend on p, T or V, and hence will not change.

Exercise 4.12

Principle: time (days) needed is inversely proportional to the *rate* of painting.

Rate, defined as the fraction of the walls painted per day, is (1/6) for person A and (1/12) for B.

Therefore the combined rate when both A and B paint will be (1/6) + (1/12) = (3/12). That is, (3/12)*th* of the walls will be painted in one day. Hence days needed to paint the walls, which is the inverse of this ratio, will be (12/3) days = 4 days.

CHAPTER 5
USE OF EQUATIONS FOR CALCULATIONS AND DEDUCTIONS

> *Which alternative below*
>
> *(statement or equation)*
>
> *is easier to memorize, recall and apply?*

Volume of gas is directly proportional to the amount of gas and its temperature and is inversely proportional to the pressure of gas.

OR

$pV=nRT$

Thinking Strategies for Solving Problems

A strategy that should always be used is the consistent and deliberate use of equations for calculations and deductions. This has many advantages. Most of the advantages stem from the fact that equations are concise, which enables them to be easily stored, recalled, processed and used.

The alternative to using equations for calculations and deductions is *verbal reasoning* (reasoning using statements). Verbal reasoning is the most common type of reasoning and it can be qualitative or quantitative. There are many types of verbal reasoning and it includes logical reasoning, direct-proportion reasoning, inverse -proportion reasoning and analogical reasoning. Verbal reasoning, however, is often cumbersome, error - prone and difficult. The difficulties are mainly associated with the need for processing lengthy information (information in verbal statements is lengthy compared with that in equations) in our short - term memory, which is very limited in most people.

Many students use verbal reasoning to solve problems that involve ratios, percentages, fractions and direct and inverse proportion relationships and they have difficulties. Instead of trying to use verbal reasoning, it will be much better to consistently use equations for all types of calculations.

The ability to store information as equations and then use them for calculations and deductions, however, needs competence in some cognitive skills. These include the skills needed to do the following operations: transform information in verbal statements into equations and vice - versa, rearrange equations and combine equations.

Mailoo Selvaratnam

The advantages of using equations for calculations and deductions will now be illustrated with some examples that have been tested with some groups of first year university students. The questions in these examples are given below to encourage you to solve them on your own, before reading through the solutions.

5.1. If three men need 12 days to perform some task, how many days will be needed by four men, working at the same rate, to do that task?

5.2. Imara has a collection of 5 cent, 10 cent and 20 cent coins. The total number of coins is 17 and the total value of the coins is 235 cents. The number of 10 cent coins is twice that of 5 cent coins. Calculate the number of coins of each type in her possession.

5.3. The mass of a fruit that has 90% water is 180 grams. It is dried to reduce the percentage of water in it to 80%. What will then be the mass of the dried fruit?

5.4. Density (d) of a substance is defined as the mass (m) of the substance divided by its volume (V).

(a) Will the density of water increase, decrease or remain unchanged if its temperature is increased?

(b) Will the density of a gas which obeys the equation $pV = kT$ (where p = pressure, V = volume, T = temperature and k is a constant) increase, decrease or remain unchanged if its temperature is increased, when the pressure of the gas is constant?

5.5. According to Graham's law, the rate of effusion of a gas is inversely proportional to the square root of the molar mass of the gas. If the rate of effusion of H_2 gas (Molar mass = 2 g mol^{-1}) is X, the rate of effusion of O_2 gas (Molar mass = 32 g mol^{-1}), under the same set of experimental conditions, will be :

(a) $(1/16)\ X$ (b) $16\ X$ (c) $0.25\ X$ (d) $4\ X$

5.6. When 2.00 moles of nitrogen and 5.00 moles of hydrogen are introduced into a closed vessel, 30 % of the nitrogen reacts to form ammonia according to the equation N_2 (g) + $3H_2$ (g) $\rightarrow 2NH_3$ (g). Calculate: (a) the amount (i.e. moles) of nitrogen reacted; (b) the amount of hydrogen reacted; (c) the amount of ammonia produced.

EXAMPLE 5.1

If three men need 12 days to perform some task, how many days will be needed by four men, working at the same rate, to do that task?

Solution

This calculation can be done by two methods: using verbal reasoning and using equations. The use of equations has many advantages and should therefore be used.

The principle involved in the calculation is the *inverse-proportion* relationship between the number of days (N_d) and the number of men (N_m) which can be represented by the equation

$$N_d = \frac{k}{N_m}$$

k is a constant. It is best to use this equation, and not verbal reasoning, to do the calculation.

The above equation has to be applied twice to do the required calculation. The first step is to calculate k using the data given (N_d = 12 when N_m = 3)

$$k = N_d \times N_m$$
$$= 12 \text{ days} \times 3 \text{ men} = 36 \text{ days men}$$

Thinking Strategies for Solving Problems

The second step is to use this value of k to calculate the required quantity (N_d when $N_m = 4$)

$$N_d = \frac{36 \text{ days men}}{4 \text{ men}} = 9 \text{ days}$$

Many groups of university students have been tested with this problem and about a third of them had difficulty. The main reason for their difficulty was their trying to manipulate the given data without first identifying the principle (inverse proportion relationship between N_d and N_m) that has to be used for doing the calculation.

EXAMPLE 5.2

Imara has a collection of 5 cent, 10 cent and 20 cent coins. The total number of coins is 17 and the total value of the coins is 235 cents. The number of 10 cent coins is twice that of 5 cent coins. Calculate the number of coins of each type in her possession.

Solution

The solution will be simplified if the three items of information given (as statements) are first transformed into equations which are then used for the calculation.

Since equations use symbols, we give symbols to the different quantities:

x = number of 5 cent coins
y = number of 10 cent coins
z = number of 20 cent coins

The three items of information given may be transformed into the following equations :

$$x + y + z = 17 \quad (1)$$
$$5x + 10y + 20z = 235 \quad (2)$$
$$y = 2x \quad (3)$$

Since the three unknowns (x, y and z) are related by the same number (three) of equations, they can be calculated by solving the equations.

Thinking Strategies for Solving Problems

Since $y = 2x$, the y term in equations (1) and (2) can be replaced by $2x$ to give respectively

$$3x + z = 17 \qquad (4)$$
$$25x + 20z = 235 \qquad (5)$$

Multiplication of both sides of equation (4) by 20 gives
$$60x + 20z = 340 \qquad (6)$$

Subtraction of equation (5) from equation (6) will give
$$35x = 105. \text{ Hence } x = 3.$$

y will then 6 by equation (3) and z will be 8 by equation (4).

EXAMPLE 5.3

The mass of a fruit that has 90 % water is 180 grams. It is dried to reduce the percentage of water in it to 80 %. What will then be the mass of the dried fruit?

Solution

Many people have difficulty with this problem. The solution will be sharpened and simplified if the following defining equation for mass percentage of the solid matter in the fruit (% solid) is used for the calculation

$$\% \text{ solid} = \frac{m_{solid}}{m_{fruit}} \times 100$$

This equation has to be applied twice. First, to calculate the quantity that remains *constant* during the drying (which is the mass of solid matter in 180g of the fruit, m_{solid}), by using the given data (% solid = 10 % because percentage of water = 90 %; m_{fruit} = 180 g)

$$m_{solid} = \frac{\% \text{ solid} \times m_{fruit}}{100} = \frac{10 \times 180g}{100} = 18g$$

This value (18 g) can then be used to calculate m_{fruit} when % solid = 20%.

$$m_{fruit} = \frac{m_{solid} \times 100}{\% \text{ solid}} = \frac{18g \times 100}{20} = 90g$$

Thinking Strategies for Solving Problems

Note that the calculation cannot be done using the defining equation for the mass percent of water in the fruit. This is because the *constant* during the drying process, m_{solid}, does not appear in the equation.

EXAMPLE 5.4

The density (d) of a substance is defined as the mass (m) of the substance divided by its volume (V).

(a) Will the density of water increase, decrease or remain unchanged if its temperature is increased?

(b) Will the density of a gas, which obeys the equation $pV = kT$ (where p = pressure, V = volume, T = temperature and k is a constant) increase, decrease or remain unchanged if its temperature is increased when the pressure of the gas is constant?

Solution

The solutions are easily deduced from the appropriate equations.

For part (a), since an increase in T will increase the volume of water (familiar experience), the equation for density, $d = m/V$, shows that d will decrease when T is increased.

Part (*b*) involves the relationship between d and T and a good method for answering this question is to first derive the equation that relates d and T and then use it for deducing the answer. The equation that relates d and T is easily derived from the two given equations ($d = m/V$ and $pV = kT$) by replacing V in the equation $d = m/V$ by kT/p (since rearrangement of $pV = kT$ shows that $V = kT/p$) and is

$$d = \frac{m\,p}{k\,T}$$

This equation shows that when T increases d will decrease (in the same proportion).

Many first year university students tested with this problem had difficulty. The main reason for this was the use of verbal reasoning. Verbal reasoning, however, is generally more difficult than using equations for making deductions.

EXAMPLE 5.5

According to Graham's law, the rate of effusion of a gas is inversely proportional to the square root of the molar mass of the gas. If the rate of effusion of H_2 gas (Molar mass = 2 g mol^{-1}) is x, the rate of effusion of O_2 gas (Molar mass = 32 g mol^{-1}), under the same experimental conditions, will be

(a) $(1/16)\ x$ (b) $16\ x$ (c) $0.25\ x$ (d) $4\ x$

Solution

The best method for solving this problem will be to give Graham's law as an equation and then use it for doing the calculation.

If the symbol R is given for the rate of effusion and M for molar mass, Graham's law may be given by the equation (k is a constant)

$$R = \frac{k}{M^{1/2}}$$

If we apply this equation to the rates of effusion of O_2 and H_2, we will get respectively

Thinking Strategies for Solving Problems

$$R_{O_2} = \frac{k}{M_{O_2}^{1/2}}$$

$$R_{H_2} = \frac{k}{M_{H_2}^{1/2}}$$

Division of the first equation by the second equation and simplification will give

$$\frac{R_{O_2}}{R_{H_2}} = \frac{M_{H_2}^{1/2}}{M_{O_2}^{1/2}} = \left(\frac{M_{H_2}}{M_{O_2}}\right)^{1/2}$$

Substitution of the data given into this equation and simplification gives

$$R_{O_2} = \left(\frac{2.0 \text{ g mol}^{-1}}{32 \text{ g mol}^{-1}}\right)^{1/2} \times x$$

$$= 0.25\, x$$

EXAMPLE 5.6

When 2.00 moles of nitrogen and 5.00 moles of hydrogen are introduced into a closed vessel, 30% of the nitrogen reacts to form ammonia according to the equation N_2 (g) + 3 H_2 (g) → 2NH_3 (g). Calculate:
(a) amount (i.e. moles) of nitrogen reacted;
(b) amount of hydrogen reacted;
(c) amount of ammonia produced.

Solution

The balanced equation for the reaction provides the information that for each molecule (and therefore for each mole) of N_2 that reacts, three molecules (and therefore three moles) of H_2 will react and that two molecules (and therefore two moles) of NH_3 will be produced. These items of information, given as statements, may be represented by the following equations

$$\frac{n_{H2} \text{ (reacted)}}{n_{N2} \text{ (reacted)}} = \frac{3}{1} \qquad (1)$$

$$\frac{n_{NH3} \text{ (produced)}}{n_{N2} \text{ (reacted)}} = \frac{2}{1} \qquad (2)$$

$$\frac{n_{NH3} \text{ (produced)}}{n_{H2} \text{ (reacted)}} = \frac{2}{3} \qquad (3)$$

Thinking Strategies for Solving Problems

(a) The defining equation for the "percentage of N_2 (reacted)" is

$$\% N_2 \text{ (reacted)} = \frac{n_{N2}(\text{reacted})}{n_{N2}(\text{initial})} \times 100$$

This equation can be rearranged and used to do the required calculation

$$n_{N2}(\text{reacted}) = \frac{\% N_2 (\text{reacted}) \times n_{N2}(\text{initial})}{100}$$

$$= \frac{30 \times 2.00 \, mol}{100} = 0.60 \, mol$$

(b) From equation (1)
n_{H2} (reacted) = 3 x n_{N2} (reacted) = 3 x 0.60 mol = 1.80 mol

(c) From equation (2)
n_{NH3} (produced) = 2 x n_{N2} (reacted) = 1.20 mol

Some students' difficulties with calculations involving balanced equations are due to the use of verbal reasoning. This type of difficulty will be avoided if the information provided by a balanced equation is first converted into mathematical equations which are then used for the calculations, as done in this example

EXERCISES

Speed = distance travelled/ time; concentration = amount / volume; density = mass/ volume; the law of conservation of mass states that the total mass does not change during any chemical reaction.

Exercise 5.1
The sum of the masses of two objects A and B is 58 grams and the difference in their masses is 12 grams. Calculate the masses of the two objects. A has the larger mass.

Exercise 5.2
A car travels from A to B at a constant speed of 50 km h^{-1}, and at 100 km h^{-1} for the return journey from B to A. Will the average speed of the car for the entire journey be the average of 50 km h^{-1} and 100 km h^{-1}? If not, what will be the average speed?

Exercise 5.3
The percentage of girls in a class is 40. Calculate the number of boys in the class if the number of girls is 20.

Thinking Strategies for Solving Problems

Exercise 5.4
The mass percentage of carbon in a compound is 25. Calculate the mass of carbon present in 60 g of this compound.

Exercise 5.5
An item is bought and then sold at a price 20 % higher than the cost price. If the selling price of the item is 24 rands, what is its cost price?

Exercise 5.6
If the time needed by 2 men to make 4 ties is 8 hours, calculate the time that will be needed by 3 men to make 9 ties.

Exercise 5.7
The temperature of a gas is increased at constant volume.
(a) Will the density of the gas then increase, decrease, or remain unchanged?
(b) Will the concentration of the gas then increase, decrease, or remain unchanged?

Exercise 5.8
Will the concentration of sugar present in an aqueous solution increase, decrease or remain unchanged if: (a) temperature of the solution is increased; (b) some water is added to it; (c) a solution of NaCl is added to it.

Exercise 5.9
Will the mass of sugar present in an aqueous solution increase, decrease or remain unchanged if: (a) the temperature of the

solution is increased; (b) some water is added to it; (c) a solution of NaCl is added to it.

Exercise 5.10
Will the concentration (c) of a gas, which obeys the equation $pV = nRT$ where R is a constant, increase, decrease or remain unchanged if its temperature is increased when: (a) the volume of the gas is kept constant; (b) the pressure of the gas is kept constant.

Exercise 5.11
If the resistance of a wire of length 10 cm and area of cross section 0.020 cm² is 0.20 ohms, calculate the resistance of 40 cm of the wire whose area of cross section is 0.010 cm². Use the following principle for doing the calculation: the resistance of a wire is directly proportional to its length and inversely proportional to its area of cross section.

Exercise 5.12
0.100 mole of nitrogen (N_2) and 0.200 mole of hydrogen (H_2) are present initially in a closed 2.0 litre vessel. 0.030 mole of nitrogen then reacts to form ammonia (NH_3) according to the equation N_2 (g) + $3H_2$(g) → $2NH_3$ (g). Calculate the amounts (moles) and the concentrations of N_2, H_2 and NH_3 in the vessel.

Exercise 5.13
10.0 grams of a gaseous substance A is present in a closed vessel at 25°C. When heated to 300°C, it partially breaks down into two gases B and C according to the equation A (g)→B (g) + C (g). If the masses of B and C produced are

respectively 3.0 grams and 2.5 grams, calculate the mass of A and the total mass of all the gases present in the vessel at 300^0 C.

Exercise 5.14
Two cars A and B are 300 km apart. Both start at the same time and travel towards one another and they meet 2 hours later. If the speed of car B is 10 km h^{-1} faster than that of car A, calculate the speed of car A.

Exercise 5.15
A boat needs 3 hours to travel, upstream a river, from A to B. The journey downstream needs 2 hours. If the water current is 5 km h^{-1}, find the speed of the boat in still water.

Mailoo Selvaratnam

SOLUTIONS TO EXERCISES

Exercise 5.1
Best to represent the information given as equations and then solve the equations

$$m_A + m_B = 58 \text{ g}$$
$$m_A - m_B = 12 \text{ g}$$

Addition of these two equations gives $2m_A = 70$ g and therefore $m_A = 35$ g. From the equation $m_A - m_B = 12$ g, it can then be seen that m_B is 23 g.

Exercise 5.2
Average speed (s) will not be 75 km h^{-1}. To calculate it, the defining equation for average speed must be used: s = total distance travelled / total time .The use of this equation will show that the average speed will be 66.7 km h^{-1}.

Exercise 5.3
Use the defining equation for percentage of girls to do the calculation. (G = girls, B = boys)

Thinking Strategies for Solving Problems

$$\% \, G = \frac{N_G}{N_G + N_B} \times 100$$

$$40 = \frac{20}{20 + N_B} \times 100$$

Hence $N_B = 30$

Exercise 5.4
Use the defining equation for mass percent of carbon (C) in the compound to do the calculation.

$$\% \, C = \frac{m_C}{m_{compound}} \times 100$$

$$25 = \frac{m_C}{60g} \times 100. \quad \text{Hence } m_c = 15g$$

Exercise 5.5
Representing the information given as an equation will simplify the solution. Let C be the cost price and S the selling price. Then, because 20 % of the cost price = (20/100) × C, the equationrelating S and C will be:

$$S = C + \frac{20}{100} \times C$$

24 Rands = C + 0.2 C.   Hence C = 20 Rands

Exercise 5.6
Proportional reasoning has to be used to do the calculation and this is best done using the following equation that incorporates the needed proportional reasoning.

$$N_{ties} = k N_{men} t \tag{1}$$

This equation incorporates the principles needed to the calculation which are that the number of ties made, N_{ties}, is directly proportional to number of men, N_{men}, employed and to the time t used for making the ties. k = constant

k needed to do the required calculation must first be calculated by applying the equation to the data given.

$$k = \frac{4\,ties}{2\,men \times 8h} = 0.25\,ties\,men^{-1}\,h^{-1}$$

$$t = \frac{N_{ties}}{k N_{men}} = \frac{9\,ties}{0.25\,ties\,men^{-1}\,h^{-1} \times 3\,men} = 12h$$

Exercise 5.7
It is best to deduce the answers from the defining equations.

(a) Equation $d = m/V$ indicates that, for a fixed mass (m) of gas, if V is kept constant d will not change

(b) Equation $c = n/V$ shows that, for a fixed amount (n) of gas, if V is kept constant c will not change

Exercise 5.8
Deduce answers using equation: $c_{sugar} = n_{sugar} / V_{solution}$.

(a) If T is increased, V will increase and hence c will decrease.

Thinking Strategies for Solving Problems

(b) Addition of any solution will increase V and hence decrease c.

Exercise 5.9
Mass of sugar in the solution will not change if its temperature is increased or if water or NaCl solution is added to it.

Exercise 5.10
(a) The defining equation $c = n/V$ shows that, for a fixed amount (n) of a gas, c will not change if V does not change.

(b) Problem involves c, T (which changes) and p (which is kept constant) of a gas. The best method for solving the problem would be to first derive the equation that relates these three quantities (c, T and p) and then use this equation for doing the required deduction. The required equation is easily derived by replacing the V term in $c = n/V$ by nRT/p (since rearrangement of $pV = nRT$ shows that $V = nRT/p$).

The equation then obtained, $c = p/RT$, shows that, when p is kept constant (R is a constant), c will decrease if T is increased.

Exercise 5.11
It is best to first transform the principles, which were given in the problem as a verbal statement, into an equation and then use it for the calculation. The equation is $R = k\,l/a$ where R = resistance, l = length, a = area and k is a constant.

To calculate R, using the equation $R = kl/a$, the value of k must be known. Hence it has to be first calculated using the given data. The value of k must then be used to calculate the required resistance, which will be found to be 160 ohms.

Exercise 5.12

It is best to transform the information provided by the balanced equation into mathematical equations and then use these equations for the calculation.

The information in the balanced equation $N_2 + 3H_2 \rightarrow 2NH_3$ may be represented by the mathematical equations

$$\frac{n_{H2}\text{ (reacted)}}{n_{N2}\text{ (reacted)}} = \frac{3}{1} \tag{1}$$

$$\frac{n_{NH3}\text{ (produced)}}{n_{N2}\text{ (reacted)}} = \frac{2}{1} \tag{2}$$

(a) From the law of conservation of matter, it follows that

$$n_{N2}\text{ (in vessel)} = n_{N2}\text{(initial)} - n_{N2}\text{ (reacted)}$$
$$= 0.100 \text{ mol} - 0.030 \text{ mol} = 0.070 \text{ mol}$$

$$n_{H2}\text{ (in vessel)} = n_{H2}\text{(initial)} - n_{H2}\text{(reacted)}$$
$$= n_{H2}\text{(initial)} - 3 \times n_{N2}\text{(reacted)}$$
$$= 0.200 \text{ mol} - 3 \times 0.030 \text{ mol} = 0.110 \text{ mol}$$

Moles of in the vessel, n_{NH3}(in vessel), is equal to the moles of produced by the reaction, n_{NH3} (produced). That is

Thinking Strategies for Solving Problems

$$n_{NH3}(\text{in vessel}) = n_{NH3}(\text{produced})$$
$$= 2 \times n_{N2}(\text{reacted}) \quad (\text{by equation 2})$$
$$= 0.060 \text{ mol}$$

Moles of N_2, H_2 and NH_3 present in the vessel are therefore respectively 0.070 mol, 0.110 mol and 0.060 mol.
(b) $c_{N2} = 0.035$ mol l^{-1}, $\quad c_{H2} = 0.055$ mol l^{-1},
$c_{NH3} = 0.030$ mol l^{-1}

Exercise 5.13

Principle for calculation is the law of conservation of mass which is best used as an equation. Since total mass before reaction (at 25°C) is equal to total mass after reaction (at 300°C), we can write

$$m_A (25°C) = m_A (300°C) + m_B (300°C) + m_C (300°C).$$

Hence

$$m_A (300°C) = m_A (25°C) - m_B (300°C) - m_C (300°C)$$
$$= 10.0 \text{ g} - 3.0 \text{ g} - 2.5 \text{ g} = 4.5 \text{ g}$$

Total mass does not change and it will be 10.0 grams.

Exercise 5.14

Give symbol x for the speed of car A. Then speed of car B will be $x + 10$. Applying the equation $s \times t = d$ to the joint speed of cars A and B we get

$[x + (x + 10)]$ km h^{-1} x 2 h = 300 km. Hence $x = 70$ km h^{-1}

Exercise 5.15
Give symbol x for the speed of the boat, in km h^{-1}, in still water.

Then speed of the boat upstream and downstream will respectively be $(x - 5)$ km h^{-1} and $(x + 5)$ km h^{-1}.

Distance travelled ($d = s\,t$) upstream by the boat will therefore be $(x - 5)$ x 3 km and distance travelled downstream will be $(x + 5)$ x 2 km.

Since these two distances are equal, $(x - 5)$ x 3 = $(x + 5)$ x 2. Hence
$x = 25$.

CHAPTER 6
PROCEEDING STEP - BY – STEP WITH THE SOLUTION

However long a journey
However difficult a task
It always involves taking
Just one step at a time.

Most problems need many steps for their solutions. It would then be unreasonable to expect to see, at the start itself as many people do, the complete way to successful solution. What can be done, and should be done, is to proceed step-by-step in a logical and systematic manner. One should always concentrate on the solution of one step at a time, without being distracted by doubts concerning the solution of subsequent steps. Proceeding step-by-step is consistent with the familiar saying

> *However long or difficult a journey, it always involves taking just one step at a time*

Proceeding step-by-step simplifies problem solving because it breaks down a complex problem into simpler problems that are easier to solve. It implies:

- The breakdown of the given problem into many simpler problems;
- The solution of these simpler problems ;
- The joining together of the solutions of the simpler problems to provide the solution to the given problem

Research suggests that many difficulties with problem solving are associated with our limited capability for *simultaneously* processing many items of information in our short-term or working-memory. Memory is of two types: long-term memory and short-term memory. Information stored in long-term memory is retrieved and processed in short-term memory whenever problems are solved. A

Thinking Strategies for Solving Problems

step-by-step procedure simplifies problem solving because it breaks down a problem into simpler problems whose solutions will not need the processing of many items of information in short-term memory at any particular time.

The general strategies considered in the previous chapters, as well as other strategies (e.g. simplification of complex problems by making assumptions or approximations, use of analogical reasoning) have generally to be linked together in a logical and systematic manner to solve most problems. For solving Quantitative problems, a four-step model or procedure may be used for logically linking the strategies. This procedure uses equations and the *crucial step in the procedure is the derivation of an equation that shows how the relevant physical quantity is related to the physical quantities in the data.*

Mailoo Selvaratnam

A FOUR-STEP MODEL (PROCEDURE) FOR PROBLEM-SOLVING

Step 1: Clarify and represent the problem clearly
In this step one should:

- Identify all the data given and the goal and distinguish them by giving different symbols;
- Organize all the relevant information (data, goal, processes) in a clear, systematic, coherent and coordinated manner, for example as tables, pictures, graphs or equations.

Step 2: Identify the required quantity and write an equation for its calculation
In this step one should:

- Identify the physical quantity that has to be calculated;
- Select the most appropriate equation (which will be called the "starting equation") for calculating it. Very often, this equation will be the defining equation for the required quantity.

Step 3: Derive an equation (the calculation equation) for calculating the required quantity
The calculation equation should have the required quantity as the only unknown. To derive this equation one should do the following:

Thinking Strategies for Solving Problems

- Rearrange the starting equation so that only the required physical quantity is present on the left side of the equation;
- Replace, step-by-step, the unknown quantities in the starting equation with known quantities. To do so, select the most appropriate equation (e.g. the defining equation) for each unknown quantity and combine it with the starting equation.

Step 4: Insert the data, in appropriate units, into the calculation equation and calculate

- Ensure that the units for the data used in the calculation are consistent with one another. It is best to use SI units;
- Deduce the units for the answer as a part of the calculation.

The step-by-step problem solving procedure outlined above will now be illustrated in some examples. Step 1 will not be needed for the solution of simple problems. The questions in the seven examples have been collected together below, to encourage you to try solving them on your own before reading through the solutions.

6.1. Three objects A, B and C can be combined in six different ways: as ABC, ACB, BAC, BCA, CAB and CBA. For four objects A, B, C and D, one way of combining them is as ABCD. Write the other ways of combining of A, B, C and D.

6.2. Use the equation $t = k/N$, where k is a constant, to calculate the value of t when $N = 4$ given that $t = 12$ when $N = 3$.

6.3. Wathsala leaves home every day in her car and travels at a constant speed of 90 km h^{-1} to pick-up Amal from his office at 5 pm, and they reach home at 5:30 pm. One day Amal finished work at 4 pm and started walking home along the road and was picked-up by Wathsala. They reached home at 5:20pm. For how long did Amal walk? Assume that the average speed of the car was 90 km h^{-1}.

6.4. A closed vessel contains an ideal gas at a pressure p and temperature T. Derive the equation that shows how the concentration (c) of this gas is related to its temperature and pressure. ($c = n/V$; ideal gases obey the equation $pV = nRT$ where n = moles, V = volume, R is a constant).

6.5. A vessel contains an ideal gas of known molar mass M at a known temperature T and pressure p. Show how the density d of this gas can then be calculated. ($d = m/V$; ideal gas obeys $pV = nRT$: $M = m/n$)

6.6. 25.0 cm^3 of water are added to 100.0 cm^3 of an HCl solution whose concentration is 1.000 mol dm^{-3}. Calculate the concentration (c) of HCl in the solution then obtained.

6.7. A sample of commercial concentrated hydrochloric acid, which is an aqueous solution, contains 36.0% (mass percent) of hydrogen chloride (HCl) and has a density 1.18 g cm^{-3}. Calculate the concentration of HCl in the solution. Molar mass of HCl is 36.5 g mol^{-1}.

EXAMPLE 6.1

Three objects A, B and C can be combined in six different ways: as ABC, ACB, BAC, BCA, CAB and CBA. For four objects A, B, C and D, one way of combining them is ABCD. Write the other ways of combining the four objects.

Solution

To ensure that no combination is missed, it is necessary to *proceed step-by-step* in a systematic manner.

1st object	Ways of arranging 2nd object	Ways of arranging 3rd object	Ways of arranging 4th object
A	B	C D	D C
	C	B D	D B
	D	B C	C B

Consider the first object in the arrangement of the four objects: it can be A, B, C or D. When A is the first object, the second object can be B, C or D and the various possible arrangements of the third and fourth objects are shown in the table below.

6 arrangements (ABCD, ABDC, ACBD, ACDB, ADBC and ADCB) are possible, as shown above, when A is the first object. Similarly 6 arrangements are possible when B,

Thinking Strategies for Solving Problems

and when C and when D, is the first object. Hence the total number of possible arrangements is 24.

A simpler problem - the combination of three objects A, B and C - has been tested with first year university students. About a third of them did not state all the 6 combinations that were possible: they identified only some of the combinations. This was mainly because they did not proceed systematically with the solution in a step-by-step manner.

EXAMPLE 6.2

Use the equation $t = k/N$, where k is a constant, to calculate the value of t when $N = 4$ given that $t = 12$ when $N = 3$.

Solution

Step 1 of the problem solving procedure is not needed because the problem is clearly stated.

Step 2: *Identify the required quantity and write an equation for its calculation*

The objective is to calculate t when $N = 4$ using the following equation (k is a constant)

$$t = \frac{k}{N} \qquad (1)$$

Step 3: *Derive an equation for calculating the required quantity*

Equation (1) cannot be used to calculate t because the value of the constant k is not known. k can, however, be calculated using the data given ($t = 12$ when $N = 3$) using the given equation after rearranging it

$$k = N \times t = 3 \times 12 = 36 \qquad (2)$$

Thinking Strategies for Solving Problems

An equation that can be used to do the required calculation can be obtained by replacing the k term in equation (1) by equation (2) when we get

$$t = \frac{36}{N} \qquad (3)$$

Step 4: *Insert the data, in appropriate units, into the calculation equation and calculate*

Using equation (3), t can be calculated when $N = 4$: the answer is 9.

About a third of the first year university students tested could not do this calculation correctly. The main reason for their difficulty was their attempting to do the calculation in one step. They did not know how to proceed with the solution using step-by-step reasoning.

This type of calculation, which involves just one equation (e.g. $c = n / V$, $s = d / t$, $F = ma$, $E = mV^2/2$, $pV = nRT$) that has to be applied twice, is tested frequently in physics and chemistry (see Example 6.6).

EXAMPLE 6.3

Wathsala leaves home every day in her car and travels at a constant speed of 90 km h^{-1} to pick-up Amal from his office at 5 pm, and they reach home at 5:30 pm. One day Amal finished work at 4 pm and started walking home along the road and was picked-up by Wathsala. They reached home at 5:20 pm. For how long did Amal walk? Assume that the speed of the car everyday was 90 km per hour.

Solution

This problem is difficult to solve unless one focuses sharply on the goal (what needs to be calculated) and does not get distracted by the irrelevant information given (speed of the car). It is also necessary to proceed step-by-step with the solution and give different symbols to the different quantities.

Focus on what needs to be calculated: the time for which Amal walked. Give it the symbol t_{walk}. To calculate t_{walk}, it will be logical to relate it to the other times that can be calculated from the given data. These are the times for which he travelled by car (t_{car}) and the total time (t_{total}). The equation relating these three times is

$$t_{total} = t_{walk} + t_{car} \qquad (1)$$

To calculate t_{walk} we must first calculate t_{total} and t_{car}.

Thinking Strategies for Solving Problems

Since Amal started walking at 4 pm and reached home at 5:20 pm

$$t_{total} = 5.20 \text{ pm} - 4 \text{ pm} = 80 \text{ minutes} \qquad (2)$$

To see how t_{car} can be calculated, see diagram below

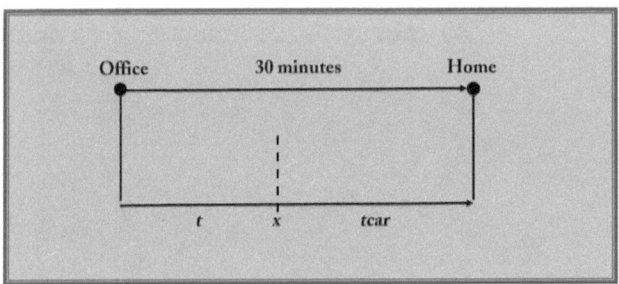

The upper line shows the time needed by the car to travel regularly from office to home, which is 30 minutes. The lower line shows that 30 minutes is equal to the sum of t and t_{car} where t is the time that would be needed by the car to travel from office to the place x on the road where Amal was picked up. From the two lines it can be seen that 30 minutes = $t + t_{car}$ and hence

$$t_{car} = 30 \text{ minutes } - t$$

Consider the calculation of t. Since the car, during its regular travel, would have to travel from x to the office and then back from the office to x (see figure), t would be equal to a half of the saving in time (which is 10 minutes; 5.30 pm - 5.20 pm). That is, t = 10 minutes / 2, which is equal to 5 minutes and therefore

$$t_{car} = 30 \text{ minutes} - 5 \text{ minutes} = 25 \text{ minutes} \qquad (3)$$

Using equations (1), (2) and (3), the required quantity, t_{walk}, can be calculated

$$\begin{aligned} t_{walk} &= t_{total} - t_{car} \\ &= 80 \text{ minutes} - 25 \text{ minutes} = 55 \text{ minutes} \end{aligned}$$

Thinking Strategies for Solving Problems

EXAMPLE 6.4

A closed vessel contains an ideal gas at a pressure p and temperature T. Derive the equation that shows how the concentration (c) of this gas is related to its temperature and pressure. ($c = n/V$; ideal gases obey the equation $pV = nRT$ where n = moles, V = volume, R is a constant).

Solution

Step 1 of the problem solving procedure is not needed because the problem is clearly stated.

Step 2: *Identify the required quantity and write an equation for its calculation*

The defining equation for concentration is

$$c = \frac{n}{V} \qquad (1)$$

Step 3 : *Derive an equation for calculating the required quantity*

The problem is to derive the equation that shows how the concentration of a gas is related to p and T. This can be done replacing V in equation (1) by nRT/p (because rearrangement of $pV = nRT$ shows that $V = nRT/p$). We then get

$$c = \frac{n}{V} = \frac{n}{nRT/p} = \frac{p}{RT} \qquad (2)$$

Equation (2) shows the required relationship between c, p and T for an ideal gas.

EXAMPLE 6.5

A vessel contains an ideal gas of known molar mass M at a known temperature T and pressure p. Show how the density d of this gas can then be calculated. ($d = m/V$; ideal gas obeys $pV = nRT$; $M = m/n$)

Solution

Step 1: of the problem solving procedure is not needed because the problem is clearly stated.

Step 2: *Identify the required quantity and write an equation for its calculation*

The problem involves calculation of density and the first step in the calculation should be the writing of the defining equation for density which is

$$d = \frac{m}{V} \qquad (1)$$

Step 3 : *Derive an equation for calculating the required quantity*

The unknown quantities in the above equation must be replaced with known quantities (M, T and p). The unknown V can be replaced by nRT/p (rearrangement of $pV = nRT$ gives $V = nRT/p$)

$$d = \frac{m}{V} = \frac{m}{nRT/p} = \frac{mp}{nRT} \qquad (2)$$

Thinking Strategies for Solving Problems

In equation (2), m and n are unknown quantities but m/n is equal to M (definition of M). Hence

$$d = \frac{Mp}{RT}$$

This equation can be used to calculate density because all the quantities needed for its calculation (M, p, R and T) are known.

Note that three steps were needed for the solution. The first step is the writing of the defining equation for density, the second step is the replacement of the unknown quantity V in the equation by the known quantities p and T, and the third step is the replacement of the unknown quantities m and n in equation (2) by the known quantity M.

Three equations were needed for the solution of the problem.

Mailoo Selvaratnam

EXAMPLE 6.6

25.0 cm³ of water are added to 100.0 cm³ of a 1.000 M HCl solution. Calculate the concentration (c) of HCl in the solution then obtained. ($c = n/V$; M denotes concentration in mol dm⁻³)

Solution

Step 1: *Clarify and represent the problem clearly*
Since two solutions of HCl are involved, confusion will be avoided if we distinguish them by giving different symbols. The symbols (i.s) and (f.s) will be used for representing the initial and final solutions. That is

$\quad$ HCl(i.s) = initial solution of HCl

$\quad$ HCl(f.s) = final solution of HCl

Step 2: *Identify the required quantity and write an equation for its calculation*
Start the solution with the defining equation for the required quantity

$$C_{HCl}(f.s) = \frac{n_{HCl}(f.s)}{V(f.s)} \qquad (1)$$

Step 3: *Derive an equation for calculating the required quantity*

Equation (1) cannot be used to calculate c_{HCl} (f.s) because n_{HCl}(f.s) and V(f.s) are not known. They should therefore be replaced with the known quantities.

V(f.s), the volume of the final solution, will be equal to the sum of the volumes of the two solutions that were mixed together. That is

$$V(f.s) = V_{water} + V_{HCl}$$
$$= 100.0 \text{ cm}^3 + 25.0 \text{ cm}^3 = 125.0 \text{ cm}^3 \quad (2)$$

n_{HCl}(f.s), the moles of HCl in the final solution, will be the same as that in the initial solution, n_{HCl}(i.s), because moles of HCl in a solution cannot change if water is added to it. Hence

$$n_{HCl}(f.s) = n_{HCl}(i.s) \quad (3)$$

Though n_{HCl} (i.s) is not known, it is related to the data given (c_{HCl}(i.s) and V(i.s)) by the equation

$$n_{HCl}(i.s) = c_{HCl}(i.s) \, V_{HCl}(i.s) \quad (4)$$

The calculation equation can now be obtained by replacing V(f.s) and n_{HCl}(f.s) in equation (1) by equations (2) and (4)

$$c_{HCl}(f.s) = \frac{c_{HCl}(i.s) \; V_{HCl}(i.s)}{125.0 \text{ cm}^3} \quad (5)$$

Step 4 : *Inset the data, in appropriate units, into the calculation equation and calculate.*

$c_{HCl}\ (i.s) = 1.000\ \text{mol dm}^{-3}$; $V_{HCl}\ (i.s) = 100.0\ \text{cm}^3$

Hence, using equation (5)

$$c_{HCl}\ (f.s) = \frac{1.000\ \text{mol dm}^{-3} \times 100.0\ \text{cm}^3}{125.0\ \text{cm}^3} = 0.800\ \text{mol dm}^{-3}$$

Comment

We solved the problem using a step-by-step procedure. A simpler method for getting equation (5) from equation (1) is to recognize that, in this problem, n_{HCl} remains constant (it does not change when water is added). c_{HCl} is therefore inversely proportional to V (volume of solution).

This implies that $c_{HCl}\ (f.s)\ V(f.s) = c_{HCl}\ (i.s)\ V(i.s)$, which is the same as equation (5).

Thinking Strategies for Solving Problems

EXAMPLE 6.7

A sample of commercial concentrated hydrochloric acid, which is an aqueous solution, contains 36.0 % (mass percent) of hydrogen chloride (HCl) and has a density 1.18 g cm^{-3}. Calculate the concentration of HCl in the solution. The molar mass of HCl is 36.5 g mol^{-1}.

Solution

Step 1: *Clarify and present the problem clearly*
We give the following symbols for the data and the required quantity.

c_{HCl} = concentration of HCl in the solution.
%$_{HCl}$ = mass percent of HCl in the solution = 36 % (data).
$d_{solution}$ = density of HCl solution = 1.18 g cm^{-3} (data).
M_{HCl} = molar mass of HCl = 36.5 g mol^{-1} (data).

Step 2: *Identify the required quantity and write an equation for its calculation*
As a general strategy, to calculate any quantity we should start with an appropriate equation for that quantity. To calculate c_{HCl} we start with its defining equation which is

$$c_{HCl} = \frac{n_{HCl}}{V_{solution}} \qquad (1)$$

Step 3: *Derive an equation for calculating the required quantity*

To calculate c_{HCl} using equation (1) we need to know the amount of HCl, n_{HCl}, present in a known volume of the solution, $V_{solution}$.

n_{HCl} is related to the molar mass of HCl, M_{HCl} (given in data), by the defining equation for molar mass ($M = m/n$):

$$n_{HCl} = \frac{m_{HCl}}{M_{HCl}} \qquad (2)$$

The unknown quantity, m_{HCl}, in the above equation is related to mass percent of HCl (% HCl; this is given in data) by the defining equation for mass percent:

$$\%HCl = (m_{HCl}/m_{solution}) \times 100 \qquad (3)$$

Equations (1), (2) and (3) can be combined to give the calculation equation (since $m_{solution}/V_{solution} = d_{solution}$)

$$c_{HCl} = \frac{\%\,HCl \times d_{solution}}{100 \times M_{HCl}} \qquad (4)$$

Step 4: *Insert the data, in appropriate units, into the calculation equation and calculate*

c_{HCl} can be obtained by substituting the given data in equation (4) and calculating

Thinking Strategies for Solving Problems

$$C_{HCl} = \frac{36 \times 1.18 \text{ g } cm^{-3}}{100 \times 36.5 \text{ g } mol^{-1}} = 1.20 \times 10^{-2} \, mol \, cm^{-3} = 12.0 \, mol \, dm^{-3}$$

Four steps were needed to solve this problem. In each step, an equation was used. The four equations, which are all defining equations, are $c = n/V$, $M = m/n$, $d = m/V$ and % HCl = $100 \times m_{HCl}/m_{solution}$. These equations are not difficult to understand and apply. Most students' difficulties are due to their not logically linking, step by step, the four steps in the solution.

Another important reason for students' difficulties is their not distinguishing clearly the different quantities in the problem and giving them different symbols. For example, many students do not distinguish between m_{HCl} and $m_{solution}$: to calculate m_{HCl} they incorrectly substitute the values for $d_{solution}$ and $V_{solution}$ in $m = d\,V$. The value of m then obtained is the mass of solution and not the mass of HCl in the solution. To prevent this type of error, symbols in equations should be explicitly labelled. Thus we should write

$$m_{solution} = d_{solution} \times V_{solution} \qquad (\text{not } m = d \times V)$$

$$n_{HCl} = m_{HCl}/M_{HCl} \qquad (\text{not } n = m/M).$$

EXERCISES

Exercise 6.1
Consider Example 6.1. State the different ways in which four objects A, B, C and D can be arranged when: (a) B is the first object; (b) C is the first object; (c) D is the first object.

Exercise 6.2
Hema invested some money in 1990. Its value doubled every 5 years. Its value in 2010 was R80 000. How much money did she invest in 1990?

Exercise 6.3

Find the total number of squares (of all sizes) present in the diagram shown above. It is not 16 because there are larger squares than the 16 smallest size squares.

Exercise 6.4
If the kinetic energy (E_k) of an object moving at a speed v is x, which one of the following will be its kinetic energy if its speed is doubled to 2 v? (Note: $E_k = mv^2 / 2$ where m = mass and v = speed)

(a) 0.5 x (b) 2 x (c) 4 x (d) x^2

Exercise 6.5
Calculate the concentration of HCl in the solution obtained when 50.0 cm^3 of water are added to 200 cm^3 of 0.100 M HCl solution (M denotes concentration in mol dm^{-3}).

Exercise 6.6
State clearly how you will prepare 100 cm^3 of a 0.100 M HCl solution starting with a 0.500 M HCl solution

Exercise 6.7
15.0 cm^3 of a 0.100 M HCl solution are added to 25.0 cm^3 of a 0.120 M NaOH solution. Calculate the concentration of NaOH in the solution then obtained. The balanced equation for the reaction is HCl + NaOH→NaCl + H_2O.

Exercise 6.8
The volume of a gas is 450 cm^3 when its temperature is 300 K and pressure is 1 atm. Calculate, using the equation $V = kT/p$ where k is a constant, the volume of the gas when its temperature and pressure are increased to 360 K and 2 atm.

Mailoo Selvaratnam

SOLUTIONS TO EXERCISES

Exercise 6.1
(a) BACD, BADC, BCAD, BCDA, BDAC, BDCA
(b) CABD, CADB, CBAD, CBDA, CDAB, CDBA
(c) DABC, DACB, DBAC, DBCA, DCAB, DCBA

Exercise 6.2
Value of the investment in 2005, 2000, 1995 and 1990 can be obtained by *halving* each time and is shown below:

2010: R 80 000; 2005: R 40 000; 2000: R 20000;

1995: R 10 000; 1990: R 5000

Exercise 6.3
Total number of squares = $4^2 + 3^2 + 2^2 + 1^2 = 16 + 9 + 4 + 1 = 30$

If the length of a side in the smallest squares is denoted by x, squares can be formed when the length is x, $2x$, $3x$ and $4x$. The number of squares formed will be 4^2 (for x), 3^2 (for $2x$), 2^2 (for $3x$) and 1^2 (for $4x$)

Exercise 6.4
To calculate E_k at any given speed v, using the equation $E_k = mv^2/2$, it is necessary to know the value for the constant m.

Thinking Strategies for Solving Problems

The first step in the calculation is therefore to obtain an expression for m using the given data:

$$m = \frac{2E_k}{v^2} = \frac{2x}{v^2}$$

Using this expression for m, the kinetic energy can be calculated when the speed is $2v$

$$E_k = \frac{2x}{v^2} \times \frac{(2v)^2}{2} = 4x$$

Exercise 6.5

$c_{HCl} = 0.080$ mol dm^{-3} (for steps in calculation, see Example 6.6)

Exercise 6.6

Focus on the goal: on the solution to be prepared which is 100 cm^3 of 0.100 mol dm^{-3} HCl. This solution contains 0.0100 mol HCl (apply equation $n = cV$) and therefore 0.0100 mol HCl must be taken from the 0.500 mol dm^{-3} HCl solution.

Volume of 0.500 mol dm^{-3} HCl solution that contains 0.0100 mol HCl is 0.0200 dm^3 (apply equation $V = n/c$). Therefore 0.0200 dm^3 (i.e. 20.0 cm^3) must be withdrawn from the 0.500 mol dm^{-3} HCl solution and diluted with water to 100 cm^3, to prepare the required solution.

Exercise 6.7

The problem involves three solutions: initial HCl solution, initial NaOH solution and final solution obtained after mixing the two initial solutions. The symbol (i.s) will be given to specify an initial solution and (f.s) to specify the final solution.

The first step in the calculation is to write the defining equation for the quantity that has to be calculated, $c_{NaOH}(f.s)$

$$c_{NaOH}(f.s) = \frac{n_{NaOH}(f.s)}{V(f.s)} \quad (1)$$

The steps in the calculation of $V(f.s)$ and $n_{NaOH}(f.s)$ are briefly indicated below

$$V(f.s) = V_{NaOH}(i.s) + V_{HCl}(i.s) = 25.0 cm^3 + 15.0 cm^3 = 40.0 cm^3 \quad (2)$$

$n_{NaOH}(f.s) = n_{NaOH}(i.s) - n_{NaOH}(reacted)$ (conservation of matter)

$\qquad = n_{NaOH}(i.s) - n_{HCl}(i.s)$ (HCl added reacts with NaOH, in equimolar amounts)

$\qquad = c_{NaOH} V_{NaOH}(i.s) - c_{HCl} V_{HCl}(i.s)$ (apply equation $n = cV$)

$\qquad = (0.120 \text{ mol dm}^{-3})(25.0 \times 10^{-3} \text{ dm}^3) - (0.100 \text{ mol dm}^{-3})(15.0 \times 10^{-3} \text{ dm}^3)$

$\qquad = 0.0015 \text{ mol} \quad (3)$

Insertion for the values for $V(f.s)$ and $n_{NaOH}(f.s)$ given by equations (2) and (3) into equation (1) and calculation will show that $c_{NaOH}(f.s) = 0.0375 \text{ mol dm}^{-3}$.

Thinking Strategies for Solving Problems

Exercise 6.8

The first step is the calculation of the constant k in the equation $V = kT/p$ using the given data:

$$k = pV/T = (1 atm)(450 cm^3)/300K = 1.50\, atm\, cm^3\, K^{-1}.$$

This value of k can then be used to calculate the required volume

$$V = kT/p = (1.50\, atm\, cm^3\, K^{-1})(360\, K)/2\, atm = 270\, cm^3$$

www.ingramcontent.com/pod-product-compliance
Lightning Source LLC
Chambersburg PA
CBHW021958170526
45157CB00003B/1048

9781482862768